THE BUTTERFLY EFFECT

THE BUTTERFLY EFFECT

Insects and the Making of
the Modern World

EDWARD D. MELILLO

ALFRED A. KNOPF • NEW YORK • 2020

THIS IS A BORZOI BOOK
PUBLISHED BY ALFRED A. KNOPF

Copyright © 2020 by Edward D. Melillo

www.aaknopf.com

Knopf, Borzoi Books, and the colophon
are registered trademarks of Penguin Random House LLC.

Grateful acknowledgment is made to the following for permission
to reprint previously published material:

Doubleday: Excerpt from "Haiku" from *Introduction to Haiku* by Harold Gould Henderson,
copyright © 1958 by Harold G. Henderson. Reprinted by permission of Doubleday, an imprint
of the Knopf Doubleday Publishing Group, a division of Penguin Random House LLC.
All rights reserved.

Grove/Atlantic, Inc.: Excerpt of "The Common Fruit Fly" from *Mount Clutter* by Sarah Lindsay,
copyright © 2002 by Sarah Lindsay. Reprinted by permission of Grove/Atlantic, Inc.
Any third party use of this material, outside of this publication, is prohibited.

W. W. Norton & Company, Inc.: Excerpt of "Touch Me" from *The Wild Braid: A Poet Reflects
on a Century in the Garden* by Stanley Kunitz Genine Lentine.
Reprinted by permission of W. W. Norton & Company, Inc.

Library of Congress Cataloging-in-Publication Data
Names: Melillo, Edward D. author.
Title: The butterfly effect : insects and the making of the modern world / Edward D. Melillo.
Description: First edition | New York : Alfred A. Knopf, 2020. | Includes index.
Identifiers: LCCN 2019053343 (print) | LCCN 2019053344 (ebook) |
ISBN 9781524733216 (hardcover) | ISBN 9781524733223 (ebook)
Subjects: LCSH: Beneficial insects. | Insects—Ecology. | Insects—Effect of human beings on.
Classification: LCC SF517 .M45 2020 (print) | LCC SF517 (ebook) | DDC 595.7/163—dc23
LC record available at https://lccn.loc.gov/2019053343
LC ebook record available at https://lccn.loc.gov/2019053344

Jacket image: Various moths and butterflies by Kubo Shunman. H. O. Havemeyer Collection,
bequest of Mrs. H. O. Havemeyer, 1929. The Metropolitan Museum of Art, New York.
Jacket design by Chip Kidd

Manufactured in the United States of America
First Edition

To my son, Simon Zev Melillo
It is a joy to watch your metamorphosis

Our treasure lies in the beehive of our knowledge. We are perpetually on the way thither, being by nature winged insects and honey gatherers of the mind.

—FRIEDRICH NIETZSCHE,
On the Genealogy of Morals: A Polemic (1887)

CONTENTS

Map   x

Introduction   3

PART I   METAMORPHOSES

1 The Bug in the System   13

2 Shellac   39

3 Silk   56

4 Cochineal   78

5 Resurgence and Resilience   96

PART II   HIVES OF MODERNITY

6 Nobel Flies   119

7 Lords of the Floral   134

8 A Six-Legged Menu   152

EPILOGUE Listening to Insects   173

Acknowledgments   179

Notes   181

Works Cited   205

Index   239

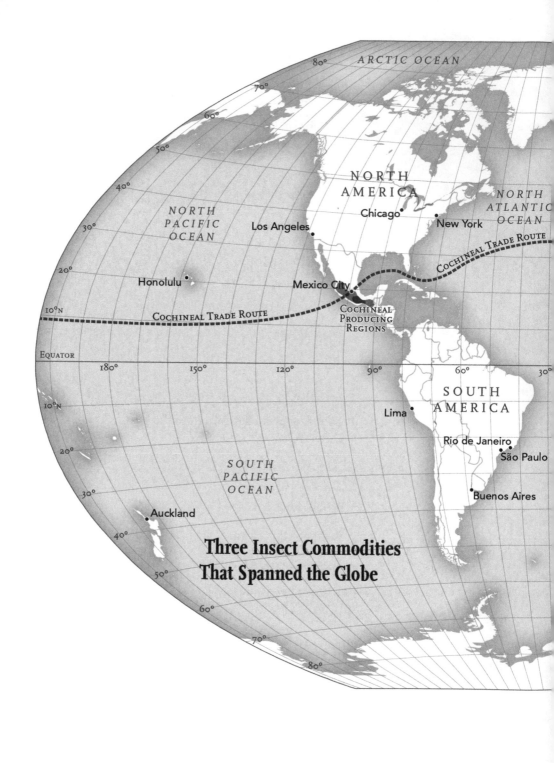

Three Insect Commodities
That Spanned the Globe

THE BUTTERFLY EFFECT

INTRODUCTION

F ear of insects is among the most prevalent anxieties of the modern age. Homeowners fret about armies of termites or legions of carpenter ants tunneling through walls and floorboards, hotel managers dread bedbug infestations, and parents live in trepidation of lice colonizing the tender scalps of their unsuspecting children. Outbreaks of mosquito-borne diseases such as Zika, West Nile virus, yellow fever, malaria, and dengue fever provide forceful reminders of the devastating effects winged pests have had on humanity.

In the realm of food production, insects play a similarly menacing role. The world's farmers spend upwards of $16 billion annually on insecticides in an endless struggle to thwart invasions of fruit flies, gypsy moths, aphids, locusts, ground beetles, and bollworms, which jeopardize harvests and threaten productivity. Despite this barrage of toxic deterrents, insects destroy an astounding 25 percent of the goods and services produced each year in developing nations. Economists have taken to measuring the impact of these creatures on a country's economic productivity as a key indicator of modernization.

Insects have often fared no better in the popular imagination. Many Western cultures equate cockroaches, flies, maggots, and fleas with filth, decay, and moral degeneration. English, like other Western languages, hums with pesky bugs: flies in the ointment, butterflies in our stomachs, ants in our pants, and bees in our bonnets. From the barfly getting buzzed at the fleabag motel to the obstinate bug in the system, predatory insects infest the vernacular environment. It is no wonder that so many Hollywood scriptwriters cast swarms of colossal creepy-crawlies as their villains.

But instead of pursuing the detrimental impacts insects (imagined and real) have had throughout history, this book will take you on a different journey. It traces the long arc of productive relationships between insects and people, revealing a fascinating array of unexpected human dependencies on our six-legged cousins. As it turns out, these tiny creatures are the living factories for many of the commodities that animate the modern world. Insects make many of the substances that pervade our daily lives: fabrics, dyes, furniture varnishes, food additives, high-tech materials, cosmetics, and pharmaceutical ingredients.

When we bite into a shiny apple or enjoy a spoonful of strawberry yogurt, listen to the resonant notes of a Stradivarius violin or watch fashion models strut down a runway, receive a dental implant or get a manicure, we are mingling with the creations of insects. The lac bugs (*Kerria lacca*),* cochineal insects (*Dactylopius coccus*), and silkworms (*Bombyx mori*) that exuded the raw materials for the shellac, red cochineal dye, and silk in these products are wonders in their own right, miniature laboratories that defy many of our expectations, expose the limits of our understanding, and reveal forgotten connections with our fellow planetary inhabitants.

In surprising and unpredictable ways, insects also sustain many of the institutions we think of as resolutely modern. Research laboratories, agribusinesses, and trailblazing start-up firms have staked their success on relationships with small winged creatures. Fruit flies (*Drosophila melanogaster*) have been crucial to the mapping of the human genome and to many other genetic breakthroughs of the past century. Bees, butterflies, beetles, and flies serve as prolific pollinators, ensuring the survival of three-fourths of the world's flowering plants and a third of the planet's food crops. Crickets, grasshoppers, and mealworms have emerged as inexpensive protein sources, essential to the future prospects of the global food supply. As all of these examples show, six-

---

* Whenever I introduce a new organism, I follow the scientific convention of providing its most up-to-date genus and species names in parentheses. This avoids the confusion of using colloquial labels, which vary among times and places.

legged creatures are as integral to the future of our earthly survival as they have been to the apocalyptic visions of pest control companies, chemical manufacturers, and screenwriters.

Invertebrates are not the only stars of this story, however. For many millennia, people in India, China, Mexico, and beyond generated the foundational knowledge on which insect-human relationships are now built. Raising domesticated insects requires an intimate awareness of these creatures' daily needs as well as a detailed understanding of how to cultivate the host plants on which these desirable bugs subsist. Rural communities on the margins of the world's political and economic power centers thus became the unofficial entomologists and the informal botanists of the Age of Discovery. From the sixteenth century onward, Europe's imperial administrators failed to comprehend the sophisticated local knowledge that sustained insect-commodity production in many parts of the world.

Colonial rivalries over these misunderstood products often generated international intrigue. In 1777, the French botanist Nicolas-Joseph Thiéry de Menonville disguised himself as an eccentric doctor, snuck into New Spain (Mexico), and absconded with hundreds of cochineal bugs. Europeans valued these insects for the dye they secrete, but Spain maintained a monopoly over cochineal production for nearly three centuries. The audacious Frenchman stowed the insects in shipping crates and smuggled them to Saint-Domingue (modern-day Haiti). The delicate creatures did not outlast Thiéry's death of "malignant fever" in 1780, but his act of biopiracy proved that cochineal cultivation had a future outside Spanish America.

This episode harked back to comparable instances of insect theft in world history. In 552 C.E., the emperor Justinian convinced two Nestorian monks to smuggle silkworm eggs from China to the Eastern Roman Empire inside their bamboo canes. The tensile threads spun by the offspring of these illicitly transported caterpillars became the basis of the Byzantine economy, which relied heavily on silk for the next 650 years. Here, as at so many other historical junctures, stories about insect lives reveal enthralling episodes in human history that would otherwise remain obscure.

Humans and insects have fashioned elaborate partnerships that outshine their history of hostile interactions. This book's title borrows from chaos theory pioneer Edward N. Lorenz, who addressed the 139th meeting of the American Association for the Advancement of Science in Boston on December 29, 1972. Lorenz's talk, "Predictability: Does the Flap of a Butterfly's Wings in Brazil Set Off a Tornado in Texas?," suggested the hallmark of chaotic behavior—namely, that small causes can have large, wide-ranging effects. His lecture spawned the term "butterfly effect." The pages that follow explore how little creatures from a single class (Insecta) of invertebrates have influenced every corner of our modern world.

Without knowing it, I have spent most of my life preparing to write this book. My connection to butterflies began when I was eleven. During the summer of 1984, I took an entomology course at the Children's School of Science in Woods Hole, Massachusetts. On the first day of class, a local naturalist named Becky Lash captivated a throng of rambunctious kids by setting a monarch butterfly (*Danaus plexippus*) on a watermelon wedge. This delicate creature proceeded to insert its proboscis into the fruit and take a long drink. Instantly, I identified with the thirst-quenching impulse of an organism that had, until that moment, seemed so stubbornly foreign.

Two decades later, I had a similar revelation about interspecies kinship. In the fall of 2003, I was living in Berkeley, California. Seeking a change of scenery, I drove down the Pacific Coast Highway to what was then the sleepy college town of Santa Cruz. While strolling along the city's oceanfront boardwalk, I noticed a flyer advertising guided tours of the Monarch Butterfly Reserve at Natural Bridges State Beach. I soon found myself following an affable young docent down a wooden footpath. She led my six-person group to a canyon where at least a hundred thousand monarchs were roosting in a sheltered eucalyptus grove. These intrepid insects had made their annual nine-hundred-mile pilgrimage from the distant Rocky Mountains to this spot, one of several dozen monarch-overwintering sites along California's coast. Clusters of velvety butterflies hung like bundles of folding fans in a bazaar, enveloping entire trees as they warmed their black, orange, and white wings in the afternoon sunshine. The spectacle was mesmerizing.

Other writers have expressed a similar blend of intimacy and astonishment at the daily lives of insects. In *Pilgrim at Tinker Creek* (1974), the American author Annie Dillard observed, "[There] are mysteries performed in broad daylight before our very eyes; we can see every detail, and yet they are still mysteries. . . . The earth devotes an overwhelming proportion of its energy to these buzzings and leaps in the grass. Theirs is the biggest wedge of the pie: Why?" As Dillard sensed, insects are among the planet's most intriguing and cosmopolitan residents. With the exception of the world's oceans, almost every ecosystem—from lush tropical forests and remote mountaintops to steamy swamps and arid deserts—is home to insects. Scientists have even discovered a species of a flightless midge that tenaciously endures Antarctica's frozen landscapes. *Belgica antarctica,* as this tiny fly is known, contains antifreeze-like body fluids that can withstand bitter frost and temperatures that plunge as low as fifty-six degrees below zero Fahrenheit (your home freezer gets down to only about five degrees Fahrenheit). Its plum-hued complexion helps this glacial survivor soak up a maximum of available sunlight.

My research for this book has also inhabited diverse realms. I draw from Sanskrit epics, Chinese folktales, Mayan legends, West African proverbs, the sixteenth-century imperial reports of Mughal emperor Akbar I, the letters of the French diamond trader Jean-Baptiste Tavernier, the Virginia journal of Captain John Smith, the stunningly rendered insect illustrations of German-born entomologist Maria Sibylla Merian, and the insect-inspired compositions of Edvard Grieg and Nikolai Rimsky-Korsakov. The annals of *Gramophone* magazine, the advertising brochures of the DuPont Chemical Company, a collection of twelfth-century Japanese short stories, and Rachel Carson's margin notes from her first draft of *Silent Spring* (1962) have also served as wellsprings of insight.

Part I of *The Butterfly Effect* explores how various cultures have come to understand our six-legged cousins over the past three millennia. Human perceptions of insects have varied widely across time and place, but many groups and individuals have found reason to temper their fear and loathing of these creatures with ample doses of admiration and wonder. Two insect products that experienced their heyday in the premodern world, honey and iron gall ink, illuminate our long-term dependency on substances created by tiny six-legged creatures.

While honey continues to serve as a fragrant and flavorful reminder of our connection to the world of bees, it is no longer a major player in the global sweetener market. Iron gall ink has virtually disappeared from all venues but museums and archives.

In contrast, I chart the enduring careers of three insect products—shellac, cochineal, and silk—that were central to indigenous human cultures in the ancient world. All of them eventually became key trade goods in European imperial economies. Finally, I chronicle the flourishing interest in these products during the mid-twentieth century and explain why they remain vital commodities in today's global economy.

The long-term histories of shellac, silk, and cochineal suggest a counternarrative to conventional accounts of modernity. During the nineteenth and twentieth centuries, industrial engineers synthesized alternatives and chemical companies aggressively promoted surrogates for all three insect secretions: vinyl for shellac, nylon for silk, and aniline dyes for cochineal. Following the Second World War, scientists, policy makers, and technocrats declared the apex of the "Synthetic Age." Yet many of these artificial substitutes for natural products turned out to be structurally inadequate or toxic to the human body. The recent reemergence of these three insect secretions as extensively traded global commodities demonstrates an enduring reliance on substances produced by domesticated insects in particular, and exemplifies a persistent dependence on "natural" substances in general.

Despite the audacious predictions of their promoters, the laboratories and factories of the Synthetic Age never achieved a monopoly on producing the material basis of everyday life. Instead, humans have (often unwittingly) relied on the productive capacity of other organisms for many of these substances. An awareness of the persistence of seemingly premodern means of production exposes cracks in the smooth edifice of technological modernity.

Part II broadens these arguments by exploring the realms of the modern world where insects have most deeply influenced human affairs. Beginning with the investigations of the Harvard entomologist Charles William Woodworth in the late 1800s and continuing with biologist Thomas Hunt Morgan's twentieth-century chromosome experiments in his Columbia University "Fly Room," the humble fruit fly has served as the preeminent model organism for genetic research.

In 2017, scientists Jeffrey C. Hall, Michael Rosbash, and Michael W. Young won the Nobel Prize in Physiology or Medicine for using *Drosophila* to understand the molecular gene responsible for controlling circadian rhythm, the physical, mental, and behavioral changes that follow a daily cycle. The fruit fly's obvious physiological remoteness from a human body conceals deep-seated genetic resemblances. Indeed, researchers have found that nearly three-quarters of human disease-causing genes have a similar structure and evolutionary origin—what geneticists call a "homologue"—in *Drosophila*.

Likewise, global agriculture would grind to a halt without the services of trillions of six-legged pollinators. By enabling the reproduction of flowering plants, wild and domesticated insects sustain massive swaths of fruits, nuts, vegetables, and feed crops like alfalfa—as well as the trees, grasses, and shrubs in rural or uninhabited regions of the world. This extraordinary coevolutionary development between bugs and blossoms keeps the planet's intricate food webs in working order.

Insects are also a source of sustenance in their own right. House crickets (*Acheta domesticus*) are a widely advertised solution to the future of global food security. Start-up firms with catchy names like Big Cricket Farms and Bugsolutely have received sizable investments from such tech-sector luminaries as Microsoft magnate Bill Gates and billionaire entrepreneur Mark Cuban. While entomophagy—the eating of insects—may strike many readers as a novel (and stomach-churning) turn of events, it is not as eccentric or innovative as it appears. The vast majority of the world's cultures have dined on a menu of six-legged creatures for millennia. Globally, at least two billion people eat them on a regular basis. In fact, researchers have cataloged more than nineteen hundred commonly consumed insect species.

The insects so deeply enmeshed in what I call "the hives of modernity"—laboratory science, agribusiness, and the food security complex—are creatures whose segmented bodies, brittle exoskeletons, and twitching antennae remind us of a prehistoric era. This paradox resides at the very core of what it means to be modern.

It is a characteristic of our era to underestimate the fundamental roles that nonhuman life-forms play in sustaining the biosphere. The Harvard biologist E. O. Wilson once remarked, "If all mankind were to disappear tomorrow, it is unlikely that a single insect would go extinct. . . . In two or three centuries, the ecosystems of the world would regenerate back to the rich state of equilibrium that existed 10,000 years ago. But if insects were to vanish, the terrestrial environment would collapse into chaos." Wilson's assertion challenges the notion that our planet is dependent on humanity; the biosphere needs insects more than it needs us.

In fact, one of the key discoveries of the modern revolution in the life sciences was the insight that an organism's scale did not necessarily reflect its evolutionary complexity or its ecological importance. As Charles Darwin noted in *The Descent of Man* (1871), "From the small size of insects, we are apt to undervalue their appearance. If we could imagine a male *Chalcosoma* with its polished, bronzed coat of mail, and vast complex horns, magnified to the size of a horse or even of a dog, it would be one of the most imposing animals in the world." In my environmental history courses at Amherst College, I never fail to remind my students that our chief local celebrity, the poet Emily Dickinson, possessed an unusual appreciation for the importance of little creatures to the grand scheme of things. As she persuasively put it, "To make a prairie it takes a clover and one bee."

Millennia of environmental adaptations and cultural accommodations have created our complex relationships to insects. Throughout history, these plucky creatures have populated our folktales, pollinated our flowers, and plagued our fields. They have served as culinary cornerstones, even as they dined on our architectural foundations. Civilizations have feared them as vectors of disease, revered them as sacred objects, and ceaselessly dissected them under laboratory microscopes.

The lives of insects offer many paradoxes. These creatures can be microscopically unobtrusive, yet they display devastating power in a swarm; it takes only one swat to destroy them, but they epitomize evolutionary resilience on earth; they are the vectors of lethal diseases, yet they produce some of the world's most durable products. Even so, insects are not merely the relics of human manipulations. Beyond botanical and mechanical metaphors lies an appreciation of the ways in which these creatures have altered—and continue to shape—the very frameworks of our existence.

. . . .

PART I

METAMORPHOSES

د

1

THE BUG IN THE SYSTEM

I n November 1944, Decca Records released a single featuring Ella Fitzgerald and the Ink Spots. "Into Each Life Some Rain Must Fall" skyrocketed to number one on the top of the Billboard charts in the United States and inaugurated a long-term collaboration between the "First Lady of Song" and the fabled record producer Milt Gabler. A century before this musical milestone, the Ottoman sultan Abdülmecid I (1839–61) founded the Hereke Imperial Carpet Manufacture to supply elaborate silk rugs for his Dolmabahçe Palace on the Bosphorus. These extravagant carpets, among the finest ever woven, featured between three and four thousand knots per square inch. Six decades earlier, on October 19, 1781, Brigadier General Charles O'Hara of His Britannic Majesty's Coldstream Guards donned his distinctive scarlet officer's coat, strode onto the battlefield at Yorktown, Virginia, and surrendered the sword of Lieutenant General Charles Cornwallis to Major General Benjamin Lincoln of the American Continental Army.

A trio of more incongruous events, spanning three centuries, is difficult to imagine, yet these episodes share an astonishing feature. They depended on the tremendous productive capacity of domesticated insects. The brittle shellac of Ella Fitzgerald's 78 rpm record, the gossamer threads woven into the sultan's silk carpets, and the crimson cochineal used to dye the brigadier general's jacket entered the circuits of global commerce as secretions from the bodies of tiny invertebrates.* Women and men in rural corners of northeastern India, the Ottoman Empire, and southern Mexico painstakingly raised the lac bugs (*Kerria lacca*), silkworms (*Bombyx mori*), and cochineal insects (*Dactylopius coccus*) that secreted the raw materials for these products.

Unwittingly, we have inherited the legacy of human-insect partnerships that yielded Ella Fitzgerald's shellac, Sultan Abdülmecid I's silk, and Brigadier General O'Hara's cochineal. Six-legged creatures have been our unshakable companions and surreptitious roommates for millennia. The average home accommodates a remarkable profusion of insects. In 2017, following a five-continent, five-year examination of residences—ranging from urban high-rises to village bungalows—California Academy of Sciences entomologist Michelle Trautwein and her colleagues concluded, "Our lives are completely mixed up with the bugs that share our homes. . . . Every home you've ever lived in, from a rural Peruvian farmhouse to a studio apartment in Paris, is teeming with tiny life."

In a related investigation, a team of scientists donned headlamps and latex gloves to comb through fifty homes in Raleigh, North Carolina. Scouring kitchen corners, crawl spaces, basements, closets, and air-conditioning vents, they discovered more than ten thousand species of insects, along with myriad spiders, centipedes, millipedes, and other arthropods. This clandestine menagerie was blithely residing alongside its unsuspecting human hosts.

While these findings intrigued some readers and spooked others, they were unsurprising to entomologists and evolutionary biologists. For the entirety of our planetary existence, we have dwelled with insects. We dine together (and, at times, on each other), we travel in tandem, and we sometimes share beds. Such relentless interactions with insects are threaded throughout the human experience. During the spring of 1748, sixteen-year-old George Washington accompanied a team of experienced wilderness surveyors as they trekked through the verdant forests of the Shenandoah Valley. The fledgling apprentice and future United States president was dismayed to find that his bed

---

* Insects are part of the large phylum of arthropods, specifically a subsection of invertebrates with a shell, joined limbs, and a segmented body. Despite popular conception, spiders, centipedes, and millipedes are technically not insects.

often consisted of nothing more than "a little straw—matted together without sheets or anything else but only one threadbare blanket with double its weight of vermin, such as lice, fleas, etc."

Although Washington's account evoked millennia of infested bedding, some of his European forebears had not regarded cohabitation with lice and fleas as a nuisance. At times, the act of hosting six-legged creatures on one's body epitomized holiness. The union of vermin and virtue was on vivid display following one of the most notorious assassinations of the Middle Ages. On December 29, 1170, four knights in the service of King Henry II of England murdered the archbishop of Canterbury, Thomas Becket, on the flagstone steps of the prelate's cathedral altar. Becket's body lay in the icy church all night. The next day, in preparation for the burial, attendants removed a profuse assortment of garments, including a mantle, a linen vestment, a lamb's-wool coat, several cloaks, a Benedictine robe, and a shirt. The innermost layer was "a tight-fitting suit of coarse hair cloth, covered on the out side with linen, the first of its kind seen in England. The innumerable vermin [i.e., lice] which had infested the dead prelate were stimulated to such activity by the cold that his hair cloth garment"—an uncomfortable shirt worn close to the skin—"boiled over with them like water simmering in a cauldron [and] the onlookers burst into alternate fits of weeping and laughter between the sorrow of having lost such a head and the joy of finding such a saint." Suitably, Becket was propelled into the afterlife on the wings of a swarm.

In the annals of Christian piety, Becket's vigorous infestation was hardly uncommon. The Scottish philosopher David Hume recounted how the Catholic saint Robert Bellarmine (1542–1621) "patiently and humbly allowed the fleas and other odious vermin to prey upon him. *We shall have heaven,* said he, *to reward us for our sufferings: But these poor creatures have nothing but the enjoyment of the present life.*" Such examples offer a new twist on the dictum "Cleanliness is next to godliness." For the righteous, the unwashed body provided a safe haven for a holy glut of six-legged creatures.

From the sacred to the profane, insects have channeled our desires. The erotic verses of Anglican cleric John Donne's Elizabethan-era poem "The Flea" illuminate this role. A young man becomes entranced by a winged bug, which suckles on his flesh and then hops over to feed on the woman of his desires. The stanzas ripen with carnal imagery: "It sucked me first, and now sucks thee / And in this flea our two bloods mingled be." The couple's bodily fluids merge within the insect. The creature's innards have become their "marriage bed . . . cloistered in these living walls of jet." As the old adage reminds us, a meal sets the table for courtship.

Whether factual or fictional, such scenarios blur the boundaries between species. We are never without insects. This habitual intimacy helps to explain why bugs have so often served as models of determination, productivity, and resilience. During the seventh century B.C.E., Japan received its ancient name Akitsushima—a hybrid of *akitsu* (dragonfly) and *shima* (island)—from Emperor Jimmu Tennō, who likened the ancient Yamato Province to a dragonfly licking its tail. In Japanese, the dragonfly came to be known as *kachimushi,* the "victory insect," because of its hunting prowess and bravery. Molded onto sword pommels, embroidered into cloaks, carved onto armored chest plates, and displayed prominently on helmets, *kachimushi* served as symbols of a samurai warrior's fortitude in battle and his serenity in domestic life.*

Insect behavior offered similar wellsprings of inspiration to the Abrahamic religions of Judaism, Christianity, and Islam. In Proverbs (6:6), the sage king Solomon advises, "Go to the ant, thou sluggard; consider her ways, and be wise," while the Qur'an (16:68–69) proclaims, "And your Lord revealed to the bee saying: Make hives in the mountains and in the trees and in what they build: Then eat of all the fruits and walk in the ways of your Lord submissively."

In more recent times, ants and bees have served as exemplars of diligent labor. Manchester, England, a center of textile manufacturing during the Industrial Revolution, features the hardworking bee on its coat of arms, and the landing outside the city's Great Hall is known as "the Bees." Similarly, the Confederation of Mozambican Business Associations gives an entrepreneur-of-the-year award, "a statue of an

---

* There is no suitable English equivalent for the Japanese term *mushi* (蟲), which makes up the second half of *kachimushi. Mushi* embraces a sweeping array of concepts, including bugs, germs, insects, and one's spirit. Its many meanings put it in stark contrast with the more scientific Japanese term for insect, *konchū* (昆虫).

An eighteenth-century Japanese samurai helmet (*kawari-kabuto*)
featuring a dragonfly (*kachimushi*) motif. Known as the "victory insect,"
the dragonfly symbolized courage in war and composure in peacetime.

ant, chosen because of this insect's reputation as a tireless worker." For
the same reasons, the U.S. Navy's Construction Battalion ("C.B.")—
also known as the "Seabees"—adopted the bee as a symbol of persistence and industriousness. The Seabees' integral role in installing
military infrastructure for Allied operations in the Pacific Theater
earned them the motto "The difficult we do now; the impossible takes
a little longer."

The beehive has even provided a convenient model for economic
theorists. The Dutch-born physician and political philosopher Bernard Mandeville (1670–1733) is remembered chiefly for his two-volume
Fable of the Bees; or, Private Vices, Publick Benefits. Published first as a
poem in 1705, and appearing as part of a more extensive book in 1714,
Mandeville's controversial proposal used the extended metaphor of the
social beehive to argue that sinful behaviors, like pride and the pursuit
of luxury, led to the general wealth and well-being of society.

Autocrats and demagogues have also relied on insect imagery.

When choosing a family emblem for the House of Bourbon, Napoléon Bonaparte rejected the regal fleur-de-lis in favor of the resourceful honeybee. Likewise, the Prussian statesman Otto von Bismarck dreamed of a similarly insectlike precision to life: "If I had to choose the form I would rather live in again, I think it would be as an ant. Just see: this little creature lives under a perfect political organization. All ants are obliged to labour, to lead a useful life; all are industrious, and perfect subordination prevails, with discipline and order."

The display of exactitude the Iron Chancellor witnessed in the ant colony extends downward from the social network to the level of individual anatomy. Insect bodies are a meticulous marriage of form and function. The word "insect" derives from the Latin term *insectum*, a literal translation of the Greek word ἔντομον (*éntomon*), meaning "cut into sections." Indeed, all insects mature from a larval or nymph stage to an adult form with a three-part body composed of a head, a thorax, and an abdomen. This three-part structure is one of their most obvious features.

In addition, they have tough and semitransparent exoskeletons that provide their bodies with a rigid external scaffolding. Such a configuration has advantages for small creatures. Like a knight's armor, it offers a protective casing against adversaries. Exoskeletons do not expand as the animal grows. In order to enlarge, an insect sheds its old carapace and develops a new one. At first, the replacement coating is malleable like wet papier-mâché, taking time to harden. Because the pull of gravity is minimal on a small body, this evolutionary strategy works. Expand an insect to the size of a human, however, and the body would soon be crushed by the sheer weight of its own shell. This rigid sheath functions because of its built-in flex points. All insects have three pairs of jointed legs, which, quite literally, define the larger arthropod phylum to which insects belong, a grouping that also includes spiders, mites, centipedes, millipedes, and crustaceans (such as crabs, lobsters, crayfish, and shrimp). This Greek term is a fusion of *arthro*, which means "joint," and *pod*, meaning "foot" or "leg."

A pair of twitching antennae are among the most recognizable features of an insect's body. These protracted sensors give their six-legged owners unprecedented mobility. Monarch butterflies use their wispy probes to dramatic effect. Anticipating the Copernican Revolution by more than 150 million years, migratory monarchs developed a heliocentric existence. Responding to cues from the sun, they use the internal rhythms in their antennae as a solar compass to guide their autumnal exodus from the northern United States and southern Canada. This built-in navigation system allows them to travel as far as twenty-five hundred miles in an astounding journey southward to their overwintering grounds in central Mexico.

An insect's movements are also guided by a set of compound eyes, which a team of Europe's leading robotics engineers has referred to as "masterpieces of integrated optics and neural design." These multifaceted orbs are made up of many tiny ommatidia, clusters of photoreceptors surrounded by support cells and pigment cells. They give insects a sensitivity to a range of wavelengths much shorter than those detectable by the human eye. At the low end of the electromagnetic spectrum, bees are able to distinguish ultraviolet light, which is invisible to humans.* Hemispheric compound eyes also give insects a nearly 360-degree view, a distinct advantage when pursuing prey or fleeing predators.

For many insects, such nimble hunting maneuvers and death-defying stunts are aerial. Most adults have two pairs of wings, which feature pliable membranous tissue and rigid veins for support. Insects are the only invertebrates that can fly. Insect wings carry their owners aloft, provide protection, attract mates, and serve as timely warnings to potential predators. Members of a family of butterflies known as Nymphalidae even use puffy veins on their top wings to assist with hearing. The extraordinary mobility and maneuverability of winged insects give them access to a vast assortment of habitats and food sources unavailable to most of Earth's animals, including our own species. A few types of insects do not have wings. The wingless insects include such groups as springtails and silverfish. Certain spiders (insects' arachnid cousins) can "balloon" by spinning a web that acts like a parachute. This allows them to surf the wind to avoid predators and shift locations.

---

* On May 23, 1993, David Blair's movie *Wax or the Discovery of Television Among the Bees* became the first film to be streamed on the Internet. Blair's cult classic is a surreal meditation on consciousness, death, and language that reimagines the world as it might be seen through a bee's eyes.

As scientific enterprises go, distinguishing humans from other beings seems like a relatively straightforward undertaking. Yet things get more complicated as we investigate such categorical divisions. Some twenty-five hundred years ago, Plato attempted to answer the question "What is a human being?" He replied, "A featherless biped." In jest, the notorious gadfly Diogenes the Cynic plucked a chicken and brought the clucking, naked specimen to Plato's Academy in Athens, announcing, "This is Plato's man!" Faced with such ridicule, Plato had no choice but to amend his definition to include the caveat "with broad flat nails."

Plato's students attempted to add a degree of precision to such fanciful descriptions. Aristotle's prodigious *Historia animalium,* composed in the fourth century B.C.E., marked the Western world's first known attempt to systematically categorize insects. The Greek philosopher's observations about these animals were precise for their time but contained errors, such as his assertion that some insects could spontaneously breed from dew, wood, putrefying mud, or dung. Four hundred years later, Roman natural philosopher Pliny the Elder (23–79 C.E.) devoted the entire eleventh book of his thirty-seven-volume *Historia naturalis* to insects. Remarking on these abundant yet minute creatures, Pliny contended, "In no one of her works has Nature more fully displayed her exhaustless ingenuity." Philemon Holland's 1601 translation of Pliny's text inaugurated the use of the term "insect" in the English language.

Holland's catchy label appeared during a phase of extraordinary innovations in the study of these creatures. Albrecht Dürer's *Stag Beetle,* painted in 1505, is one of the most widely emulated nature studies in European history. This watercolor by the influential German artist became iconic for its detailed rendition of an imposing beetle, set against a blank background. In fact, practitioners of the early modern "nature study"—which focused on individual specimens, removed entirely from their environments—faithfully imitated Dürer's techniques.

At the other end of the century, the Dutch spectacle makers Zacharias Jansen and his father, Hans, designed and built the first compound microscope during the 1590s. By placing convex glass lenses at the top and the bottom of an opaque tube, the two craftsmen found that they could make the objects viewed through their device appear between three and nine times larger. This invention provoked a visual revolution in the sciences. Researchers could now access miniature worlds that had been previously hidden from view. Insects were often the favorite newfound objects of scrutiny. In fact, the earliest microscopes were known as "flea glasses" because of their utility for studying tiny creatures.

The microscope brought clarity to the hitherto obscure realm of insect anatomy. In 1669, the Italian biologist Marcello Malpighi used this new tool to complete the first systematic dissection of an insect. Given the minuscule dimensions of the silkworm he eviscerated and the primitive state of seventeenth-century microscopy, this was no mean feat. Many months of arduous research took their toll on Malpighi's health but yielded stunning results. As he recalled, "I was plagued next autumn with fevers and inflammation of the eyes. Nevertheless, such was my delight in the work, so many unsuspected wonders of nature revealing themselves to me, that I cannot tell it in words." Thirty years after of Malpighi's blinding revelations, an extraordinary Dutch scientist, artist, and explorer crossed the Atlantic in search of undiscovered insect worlds. Maria Sibylla Merian is not a household name. It should be. Merian is one of the founders of modern entomology.

In 1699, fifty-two-year-old Maria and her twenty-one-year-old daughter, Dorothea, crossed a gangplank at a bustling Amsterdam dock, boarded a hulking merchant ship, and sailed across the Atlantic to Surinam. Maria funded the entire expedition with the sale of some 225 of her own paintings, and their trip constituted the first purely scientific expedition to the Americas by private individuals. As two women traveling abroad unaccompanied and unsponsored, the pair flouted social norms. They left behind their comfortable middle-class existence in the Netherlands to spend the next two years roaming the tropical rain forests of the Dutch colony, where they drew, painted, and cataloged an immense array of insects and plants in their native habitats.

The Republic of Suriname—as it is now known—is a knuckle on the bulging fist of South America, tucked between Guyana and French Guiana on the continent's northeast coast. When Dutch colonial rule began in 1667, Surinam became part of a cluster of colonies called

Dutch Guiana.* While Surinam's topography was extremely familiar to its indigenous Arawak, Tiriyó, Wayana, Carib, and Akurio residents, it was a mysterious blank spot on seventeenth-century European maps. Colonizers filled this void with dreams of coveted tropical commodities like sugar, coffee, cacao, and cotton. To satisfy this economic fantasy, the Dutch West India Company imported thousands of African slaves, who lived under conditions of extraordinary cruelty. As the historian Charles Ralph Boxer remarked, "Man's inhumanity to man just about reached its limits in Surinam." The use of extreme forms of torture on slaves who failed to meet work quotas, extended public lashings of those whom masters perceived to be insubordinate, and the imposition of the death penalty for running away were among the most infamous features of this brutal system.

In stark contrast to this unchecked ruthlessness was the sheer beauty of the colony's environment. Surinam's tropical climate and its vast swaths of uninterrupted forest make it one of the world's biodiversity hot spots, an ecological wonderland teeming with life. From the sixteenth century onward, accounts of this otherworldly nature had filtered back to Europe. Tales of iridescent giant blue morpho butterflies, woolly gray moths with avian wingspans, and verdigris-tinted beetles the size of silver dollars had captivated Maria's imagination long before she set out across the Atlantic. Merian was one of the first in a long line of foreign scientists drawn to Surinam's insects. Conservation biologist E. O. Wilson opened his 1984 memoir, *Biophilia,* with a colorful description of how a visit to Surinam profoundly shaped the early stages of his career.

Initially, the two women pursued caterpillars and winged quarry among the cassava shrubs and stocky pineapples that grew "as easily as weeds" in Paramaribo's domestic gardens. In 1700, the colonial capital was home to some seven hundred Europeans. One of Merian's contemporaries described the town as "the meeting place where planters come together, where the Lords of the Government give their orders, and where the populace receives them. Here the ships sail hither and thither." The toffee-colored cones of unrefined sugar that packed the

---

* In 1978, three years after national independence, the postcolonial government changed the official spelling of the country's English name from "Surinam" to "Suriname."

holds of the departing vessels had been shaped by the eighty-five hundred enslaved Africans who labored in Surinam's sugarcane fields. Over the coming century, the ranks of this coerced workforce would swell by more than 600 percent. Discontent with the intolerable conditions on the plantations frequently boiled over into rebellion. Maroon communities of escaped slaves thrived in the forests just beyond the reach of the Dutch colonial frontier. These renegades periodically raided the settlements of their former European masters for guns and supplies.

As Maria and Dorothea ventured beyond the city limits, it became obvious that this tense borderland between freedom and slavery was also home to a vast array of biological surprises. Accompanied by African slaves and Amerindian assistants, the two women paddled a dugout canoe up the Surinam River, sampling and sketching the remarkable flora and fauna they encountered. The lure of terra incognita drew the Merians deeper into the jungle. Maria recounted, "The forest grew together so closely with thistles and thorns, I sent slaves with hatchets ahead, so they chopped an opening for me, in order to go through to some extent, which was nevertheless rather cumbersome." The jungle echoed with a symphony of notes unfamiliar to European ears. The falsetto chatter of spider monkeys mingled with the resonant squawks of white-throated toucans. A towering canopy overhead, exceeding 160 feet at times, must have confounded Maria and Dorothea's sense of proportion. Beneath a tangle of vines and snaking branches of trees with fairy-tale names like "sweet bean" and "marmalade box," the two women encountered tawny, saucer-sized tarantulas called goliath birdeaters (*Theraphosa blondi*). The world's largest arachnids emerge at dusk to make evening meals of unlucky earthworms, toads, lizards, or mice. Indeed, the nickname "birdeater" originated from Maria's portrait of this mammoth spider devouring a hummingbird.

The Merians sketched or painted more than ninety species of animals and sixty species of plants. Insects were the focal points of their studies. The Merians' illustrations vividly portray crimson caterpillars munching on glistening leaves, delicate moths laying bundles of pearly eggs, and agile butterflies sucking nectar from the pollen tubes of orchids. Eventually, Maria and her daughter transferred these images from their leather-bound study journals to membranous sheets of vellum, a fine parchment made of calfskin.

Surinam's tropical heat and sundry fauna took a toll on Maria's health. Among her less auspicious encounters with the insect world was the bite she received from a malarial *Anopheles* mosquito. As she wrote to fellow naturalist Johann Georg Volkamer, "I almost had to pay for it with my life." She began to suffer from debilitating bouts of fever, and the Merians' plans for a five-year trip were cut short. Maria and Dorothea departed Surinam on June 18, 1701, taking with them a veritable museum of discoveries. Their luggage contained jars of delicate butterflies preserved in brandy; bottles stuffed full of snakes, iguanas, and a small crocodile; cases of swollen flower bulbs and puffy chrysalids; and round boxes that cradled thousands of meticulously pressed insects. Not all of these samples were destined for research. Merian was an astute entrepreneur who funded her scientific career by selling artwork and exotic specimens to private collectors for their cabinets of curiosities.

Four years after she returned to the Netherlands, Maria published her masterwork, *Metamorphosis insectorum Surinamensium* ("Metamorphosis of the Insects of Surinam"). The volume, lavishly illustrated with sixty copperplate engravings, begins with a clear-eyed declaration of its author's approach:

> I created the first classification for all the insects which had chrysalises, the daytime butterflies and the nighttime moths. The second classification is that of the maggots, worms, flies, and bees. I retained the indigenous names of the plants, because they were still in use in America by both the locals and the Indians.

Maria's choice to preserve Native American terminology was an attempt to advertise the exotic nature of her findings, but it was also a conscious decision to maintain traditional ecological knowledge that was being ignored by newcomers in many colonial settings.

By all accounts, Merian's book was groundbreaking. Not only did the illustrations from her unusual endeavors dazzle viewers with their brilliant colors and astonishing level of detail; these pictures also broke with tradition by showing all the stages of an insect's metamorphosis

A portrait of the entomologist Maria Sibylla Merian (1647–1717) on the German
five-hundred-deutsche-mark note. At age fifty-two, Maria and her daughter
traveled to the Dutch colony of Surinam, where they spent two years drawing
and collecting insects. Maria's illustrations, which she published in 1705 as part of
her *Metamorphosis insectorum Surinamensium* ("Metamorphosis of the Insects
of Surinam"), revolutionized the understanding of insects and their life stages.

in a single image.* Furthermore, Maria's innovative compositions displayed the entire life cycle of an insect against the backdrop of its host
plant. Previously, most of her male colleagues had depicted these creatures as isolated subjects, fixed in time and lacking ecological context.
Modern-day entomologists who have analyzed Merian's images report
being able to determine the genus of 73 percent of her butterflies and
moths and the exact species of 66 percent of them.

The scientists who followed in Merian's indomitable footsteps discovered that insects have brains, hearts, digestive tracts, reproductive
organs, muscles, and nerve cells that function in ways comparable to
those of humans. Researchers have also demonstrated remarkable
likenesses in human and insect behavior. Honeybees and ants display
behavioral patterns that provide analogies to human personalities.
Some "scout" bees are thrill-seeking extroverts, while others are less

---

* Merian's Dutch contemporary Jan Swammerdam (1637–1680) also advanced human
understanding of insect life stages—egg, larva, pupa, and adult. His work helped dispel
the prevailing idea that different phases of insect development (e.g., caterpillar and butterfly) were unique organisms.

adventuresome homebodies that stay closer to the hive. Similarly, the termite-hunting ant *Megaponera analis* rescues injured coworkers, carrying them back to the nest, where they can recover. This altruistic behavior dramatically reduces combat mortalities and preserves the colony's social structure.

Despite sharing such conspicuous attributes, insects and humans had very distinctive histories. Insects evolved around 480 million years ago, long predating the comparatively recent human presence on Earth. The earliest fossil evidence for anatomically modern *Homo sapiens*—discovered on a tree-speckled savannah in 2017 at Jebel Irhoud, Morocco—is a mere three hundred thousand years old.

Insects' small stature is a relatively new phenomenon in the larger sweep of their planetary existence. Gargantuan dragonfly-like griffin-flies (*Meganeura monyi*) flourished roughly three hundred million years ago. Their two- to three-foot wingspans made them the size of small hawks. Equipped with formidable jaws and spiny front legs, these carnivorous creatures were undoubtedly an intimidating presence in prehistoric skies. The evolution of predatory birds contributed to their decline, as did reductions in atmospheric oxygen levels.

In fact, breathing has been a major limitation to insect size. Instead of relying on a circulatory system to deliver oxygen to their cells, as vertebrates do, insects respire through pairs of holes known as spiracles. These openings connect to a labyrinth of tiny tracheal tubes. Most insects passively oxygenate their cells via simple diffusion, which means that every cell in an insect's body must be adjacent to tracheae in order to receive oxygen. As a result, a substantial portion of an insect's anatomy is dedicated to housing microscopic air tubes. For example, the tracheal system occupies 39 percent of the body volume of June bugs from the insect genus *Phyllophaga*, those ungainly beetles that colonize porch lights, stagger through backyard barbecues, and cling to screen doors on hot summer nights.

Conversely, small size has offered numerous advantages to insects. Most mature to reproductive age quickly, enabling rapid population growth and swift adaptations to new or changing environments. As anyone who has experienced the unpleasant discovery of a wardrobe infested with moth larvae can confirm, bugs thrive in a broad variety of ecological niches that favor tiny bodies. In the words of the eminent

Russian biologist Sergei Chetverikov, "by continuously diminishing the dimension of their body" insects achieved "an endless variation of forms and thereby acquired a tremendous importance in the general economy of nature. Thus their insignificance became their power."

The success of this adaptive strategy is reflected in the sheer numbers of insects on our planet. "A single hectare [about two and a half acres] in the Amazon basin contains more ants than the entire human population of New York City," notes entomologist Mark Moffett. Scientists think that at any given moment, a mind-boggling ten quintillion (10,000,000,000,000,000,000) insects are alive on Earth. This means that, as of 2020, there are 1.3 billion insects for every human on the planet. In 1833, eminent British entomologist John Obadiah Westwood published the first numerical assessment of the number of Earth's insect species: "If we say 400,000, we shall, perhaps, not be very wide of the truth." Estimates now suggest an average of 5.5 million separate insect species.

Insects are among the planet's most abundant terrestrial life-forms. They make up 80 percent of Earth's animals. A single family of beetles, the weevils (Curculionidae), contains an astounding 60,000 species, more than all of the species of mammals, fish, reptiles, and birds combined. British-born Indian biologist J.B.S. Haldane (1892–1964) once found himself in the company of several theologians, who asked him what one could conclude about the nature of the Creator from a study of life on Earth. Haldane allegedly answered, "An inordinate fondness for beetles."

Such tremendous variety is matched by astonishing adaptability. The petroleum fly (*Helaeomyia petrolei*) subsists in southern California's naturally occurring oil seeps. Its larvae feed on dead insects that become trapped in the viscous asphalt of primordial tar pits. Thousands of miles across the Pacific, the Hawaiian wēkiu bug (*Nysius wekiuicola*) has engineered a habitat among the exceedingly inhospitable conditions atop the dormant Mauna Kea volcano. This tiny insect consumes other minuscule creatures blown to their deaths on the icy summit at 13,800 feet above sea level. Insects have even completed long-distance oceanic crossings. In October 1988, scientists watched in astonishment as a cloud of desert locusts survived a five-thousand-mile transatlantic journey from West Africa to the Caribbean.

Among the planet's hardiest organisms is the Hawaiian wēkiu bug
(*Nysius wekiuicola*), which survives atop the often-snowy summit
of the dormant Mauna Kea volcano by eating other tiny creatures that
die in the icy gusts at 13,800 feet above sea level.

Yet these evolutionary success stories have not brought universal
acclaim to insects. In Euro-American culture, insects traditionally
occupied the lowest tier of animal life in the Great Chain of Being, a
hierarchical understanding of the natural world that endured from the
fifth century B.C.E. to the late 1800s. Being compared to an insect has
often been a mark of misery. In the winter of 1849, a correspondent for
The Illustrated London News reported finding desperate victims of the
Irish potato famine living in a shallow, covered pit: "It was roofed over
with sticks and pieces of turf, laid in the shape of an inverted saucer. It
resembles, though not quite so large, one of the ant-hills of the African
forests." This was not the first or the last time that insects would serve
as symbols of disparagement.

During the twentieth century, a few hundred species of
"bloodsuckers"—primarily mosquitoes, ticks, flies, lice, and fleas—
dominated written accounts of how insects shaped world history.
American physician Hans Zinsser's *Rats, Lice and History* (1935) inaugurated this genre by surveying the effects of insect-borne epidemics
on human affairs. Zinsser focused on "these ferocious little fellow creatures, which lurk in the dark corners and stalk us." British naturalist
John Cloudsley-Thompson's *Insects and History* (1976) followed suit,

concentrating almost entirely on insects' destructive impacts, which ranged from ancient plagues to more contemporary outbreaks of malaria, typhus, typhoid, and yellow fever.

More recently, in *Mosquito Empires* (2010), historian John McNeill demonstrated that insects profoundly altered the early modern geo-political environment. Between the 1600s and the First World War, *Anopheles* and *Aedes* mosquitoes—highly mobile carriers of malaria and yellow fever—devastated invading armies, constrained imperial expansion, and bolstered the anti-colonial revolutions of the coastal Americas. Historian Timothy C. Winegard's 2019 book, *The Mosquito: A Human History of Our Deadliest Predator,* expands this argument by fastidiously cataloging the widespread death and suffering that mosquito-borne diseases have wrought throughout history.

Insect-borne maladies continue to shape human experience. More than one hundred species of *Anopheles* mosquitoes can transmit *Plasmodium* parasites, which cause malaria in humans. In severe cases, malaria produces fever, chills, respiratory distress, and organ dysfunction; it can be fatal if untreated. The World Health Organization estimates that 435,000 people died of malaria in 2017, most of them in Africa. Meanwhile, the *Aedes aegypti* and *Aedes albopictus* mosquitoes that carry the Zika virus were vectors for the devastating outbreak in 2015–16 that caused birth defects and neurological disorders in residents of sixty countries.

In addition to the havoc that insects have wreaked as disease carriers, they have also served as potent weapons for human armies. The Popol Vuh, or Mayan book of creation, recorded how the pre-Columbian K'iche' people of Central America released swarms of wasps on enemies who attacked their mountain citadel at Hacauitz (in today's highland Guatemala). Centuries later, in North America, many Confederates alleged that Union troops deliberately imported the orange-and-black Mexican harlequin bug (*Murgantia histrionica*) to devour southern crops. Similarly, during the Japanese occupation of Manchuria in the 1930s and 1940s, the notoriously secretive Unit 731 attempted to weaken local resistance by dropping packets of plague-infested fleas over northern China.

Yet even in the heat of combat, insects have come to our rescue. During Napoléon Bonaparte's Syrian campaign of 1799, the general's doctor observed that local fly larvae accelerated the treatment of soldiers' battlefield wounds by consuming dead tissue without harming living flesh. In his *Memoirs of Military Surgery*, Napoléon's battlefield surgeon Dominique Jean Larrey remarked, "Although these insects were troublesome, they expedited the healing of the wounds by shortening the work of nature, and causing the sloughs to fall off." Today, forensic entomologists have inherited the tool kit of eighteenth-century field medics. Modern crime scene investigators can assess the developmental stages of carrion insects to estimate the time elapsed between a victim's death and the body's discovery.

In other cases, insects have provided the decisive stimulus for economic change. The residents of Enterprise, Alabama, credit their long-term financial success to the boll weevil (*Anthonomus grandis*). For most of the twentieth century, this cotton-eating beetle wreaked havoc throughout the United States and Latin America, relentlessly devouring fields of the cash crop. However, the discerning farmers of Enterprise diversified their fields in 1919, planting peanuts, potatoes, sugarcane, and sorghum just prior to the apex of the boll weevil infestation. That year, to celebrate the profits from their expanded agricultural repertoire, the residents of Enterprise acquired a thirteen-foot-high marble statue of a robed Greek woman holding a fifty-pound iron boll weevil above her head. The benefits of bugs emerge in unexpected ways.

But when did insects become bugs? In the entomologist's lexicon, "true bugs" are insects belonging to the taxonomic order Hemiptera. The species in this group—including aphids and cicadas, along with the suggestively named "leafhoppers," "plant-hoppers," and "shield bugs"—have piercing-sucking mouthparts and specialized forewings called hemelytra. In everyday usage, however, "bug" and "insect" are interchangeable. The English word "bug" comes from the Arabic term *baq* or *bakk* (بَقّ), a colloquial Levantine name for bedbugs. According to the thirteenth-century Islamic jurist Ibn Khallikān, "There are three *b*'s which torment us, *bakk* (bugs), *burguth* (fleas), and *barghash* (gnats); the three fiercest species of created beings, and I know not which is the worst." Contemporary British English preserves these distinctions. In the United Kingdom, "bug" often means "bedbug."

The term's connotations expanded during the Industrial Revolution. Thomas Alva Edison was among the earliest engineers to use

Computer programming pioneer Grace Murray Hopper (1906–1992) discovered
a moth trapped in an electromechanical relay of the Harvard University
mainframe, helping to popularize the expression "bug in the system."

the word "bug" to describe a glitch in his electronic hardware. As the
prolific designer wrote to an associate in 1878, "The first step is an intu-
ition, and comes with a burst, then difficulties arise—this thing gives
out and [it is] then that 'Bugs'—as such little faults and difficulties are
called—show themselves and months of intense watching, study and
labor are requisite before commercial success or failure is certainly
reached." Here, as in so many other cases, insects burrowed into the
realm of metaphor.

It took more than half a century for the expression "bug in the system" to achieve widespread use. In the 1940s, the American computer
programmer Grace Murray Hopper traced a malfunction in the Harvard University mainframe to a moth trapped in an electromechanical
relay. Hopper's notebook, with the deceased offender still taped next to
her entry for the day, reads: "First actual case of bug being found." In
this case, "de-bugging" was a quite literal description.

At other times, bugs have assumed truly menacing dimensions in
the popular imagination. The modern Euro-American attitude toward
these creatures generally remains one of revulsion. In his treatise *British Central Africa* (1897), the botanist and colonial administrator Sir

Harry Johnston admitted to his "sweeping hatred of the insect race" and remarked, "It is surprising to my thinking that our asylums are not mainly filled with entomologists driven to *dementia* by the study of this horrible class." Three decades later, the popular U.S. monthly *Modern Mechanix Magazine* advised its readers, "A world ruled by giant insects, with the last remnants of the human race as slaves is one of the favorite devices of one school of fiction writers. Fantastic? Not at all. . . . In fact, all past history indicates that when, and if, the present civilization comes to an end, it will die because of an unsolved food problem, and that insects will be a contributing factor, and hence may be the survivors." Apocalyptic scenarios, set into motion by swarms of six-legged creatures, recur throughout Western culture.

Unfortunately, human disgust for insects has also been associated with some of the most horrific crimes in recorded history. Genocidal leaders and their minions often dehumanize and disparage victims by referring to them as insects because they see these populations as sub-human creatures deserving annihilation. During the California Indian genocide of 1846–73, killers sometimes referred to California Indians as lice and called their children nits. Similarly, Heinrich Himmler, the leader of Nazi Germany's Waffen-SS, proclaimed in April 1943: "Anti-semitism is exactly the same as delousing. Getting rid of lice is not a matter of ideology. It is a matter of cleanliness." Himmler, one of the principal architects of the Holocaust, validated the "Final Solution" by equating Jews with loathsome insects.

The theatrical world has also embraced the role of the insect as vile adversary. Nobel Prize–winning Belgian playwright Maurice Maeter-linck (1862–1949) remarked, "Something in the insects, however, seems to be alien to the habits, morals, and psychology of this world, as if it had come from some other planet, more monstrous, more ener-getic, more insensate, more atrocious, more infernal than our own." Movies such as *Tarantula* (1955), *The Deadly Mantis* (1957), *The Fly* (1958), *Killer Bees* (1974), *Empire of the Ants* (1977), *The Swarm* (1978), *Arachnophobia* (1990), *Mimic* (1997), *Eight Legged Freaks* (2002), *Black Swarm* (2007), *The Hive* (2008), and *Infestation* (2009) rely on stock portrayals of arachnids and insects as "evil arthropods" that wreak havoc on humankind. Hollywood has even adopted biomimicry—the imitation of nonhuman systems to solve human problems—as a strategy for transforming insects into villains. The insidious Xenomorph in the *Alien* films (1979–2017) has a life cycle modeled on parasitic wasps that lay their eggs inside living host caterpillars. As *Alien's* screenwriter Dan O'Bannon explained in 2003, "Works of fiction weren't my only sources. I also patterned the Alien's life cycle on real-life parasites. Parasitic wasps treat caterpillars in an altogether revolting manner, the study of which I recommend to anyone tired of having good dreams."

When screenwriters reverse directions and venerate insects, they typically humanize the bodies of these creatures. In Pixar/Disney's 1998 computer-animated film *A Bug's Life*—a tale inspired by Aesop's fable "The Ant and the Grasshopper"—an ant named Flik recruits a band of traveling circus bugs in an attempt to save his colony from a gang of evil grasshoppers. Screenwriter and codirector Andrew Stanton recalled his conversations about insect anatomy with the Pixar Animation Studios art department. The design team spent hours discussing the physical attributes of the virtuous ants and the malevolent grasshoppers. In the end, they decided to sacrifice realism for bodily modifications that humanized the story's heroes: "We took out mandibles and hairy segmentation yet still tried to keep design qualities and aspects of texture that made you feel like you were looking at

In *Pinocchio* (1940), Walt Disney made Jiminy Cricket more likable
to viewers by smoothing his features, removing his antennae and wings,
making him a biped, and dressing him in formal human attire.

bugs. We wanted people to like these characters and not be grossed out by them." The evil grasshoppers appeared with six appendages, barbed exoskeletons, and highly visible wings. In contrast, the artists designed ants that walked upright on two legs and had a pair of arms and smooth features. The forerunner of such "civilized" cinematic insects was Walt Disney's Jiminy Cricket in *Pinocchio* (1940), a dapper character attired in a tailcoat, top hat, opera gloves, and patent leather shoes with yellow spats.

Insects have shaped human history in profoundly material ways as well. While creepy aliens and affable bugs inhabit the silver screen and dwell on the written page, actually existing six-legged bugs have lived among us since the dawn of humanity, creating substances that animated civilizations and offered new pathways for human expression.

Insects have long served as gatekeepers to the experience of sweetness. By pollinating fruit trees and producing honey, bees quench the innate cravings for sugar common to so many animals. Humans have gathered wild honey since the Stone Age. The oldest known depictions of honey hunting appear in an eight-thousand-year-old painting on the walls of the Cuevas de la Araña (known in English as the Spider Caves) in Valencia, Spain. A graceful line drawing in crimson ink portrays two figures ascending ropes along a cliff face to gather their sticky harvest from a secluded nest. A swarm of bees buzzes nearby. Similar ancient images discovered in southern Africa attest to the widespread human pursuit of honey in its most unadulterated form.

Many millennia ago, North Africans domesticated honeybees, from the genus *Apis*. Beekeeping, or apiculture, dates at least to the reign of the ancient Egyptian pharaoh Nyuserre Ini (2474–2444 B.C.E.). An illustration in his ornate sun temple near the banks of the Nile depicts workers blowing smoke into hives in order to calm the bees before removing their honeycombs. Subsequently, apiculture flourished throughout the Mediterranean region. Beekeeping was so common in ancient Greece that some city-states subjected it to regulation. According to the Greek essayist Plutarch, the Athenian statesman Solon (c. 638–558 B.C.E.) declared, "He that would raise stocks of bees was not to place them within three hundred feet of those which another had already raised."

Honey continued to be a prime concern of kings, queens, nobles, and commoners in medieval Europe. England's Charter of the Forest, a thirteenth-century document that reestablished rights of access to royal lands, granted a freeman the right to "the honey that is found within his woods." The privilege of honey hunting was among the hallmarks of individual autonomy.

Beeswax was also a coveted material in medieval Europe. This hard yet pliable substance frequently served as currency. From the thirteenth through the sixteenth centuries, it was common for villagers and city dwellers in Russia, Germany, England, and Scotland to pay their rent in lumps of beeswax. Beeswax never spoiled, and it had countless applications. Medieval artisans used it to seal ships' hulls, coat furniture surfaces, manufacture cosmetics, and create molds for casting metals and ceramics. Most importantly, it illuminated the night. Beeswax has a high melting point of 146 degrees Fahrenheit. Candles made from it remain upright when they are lit. The oldest intact beeswax candles ever found north of the Alps were among the trove of burial objects unearthed in a graveyard on Germany's Upper Rhine River. They date to the sixth or seventh century C.E.

Beekeeping evolved independently in the Americas. Long before Europeans crossed the Atlantic, Mayas and Aztecs used hollow logs as hives for raising native stingless bees (*Melipona beecheii*). These advanced Mesoamerican civilizations put *Melipona* honey and wax to a vast range of medicinal, culinary, and religious uses. The Codex Mendoza, a detailed record of Aztec culture created in the decades following the Spanish invasion of Mexico in 1519, mentions hundreds of "*cantarillos de miel de abeja*" ("small jars of bees' honey") among the tribute Aztec emperor Montezuma II received from his rural subjects. Conquistadors dramatically expanded upon this precedent. From 1549 to 1551, Spaniards appropriated nearly twenty-five metric tons of beeswax and more than twenty-three metric tons of honey from 173 towns on the Yucatán Peninsula.

Throughout the world, countless societies depended on honey as a sweetener, a medicine, and a ceremonial element. In the northeast corner of Australia, the Yolngu people of Arnhem Land have spent millennia honing the practice of hunting for the hives of wild stingless bees (*Tetragonula carbonaria*). The expansive Yolngu term "sugarbag" refers not only to the honey and the bees in these hives but also their wax and larvae. For the successful hunter, a meal of sugarbag provides a sweet, fatty, protein-rich feast. Beyond its nutritional value, the sugarbag carries deep symbolic meaning that permeates songs, dances, kinship ties, ancestral connections, and sacred objects.

In other contexts, honey also gave humans access to new realms of consciousness. Mead, the fermented alcoholic beverage brewed from water, honey, and naturally present yeasts, appears in the folklore of so many cultures that its origins are murky. What is clear, however, is its central role in antiquity. *Mádhu,* the Sanskrit root for the English word "mead," occurs more than three hundred times in the Rigveda, a collection of Vedic hymns composed sometime between 1500 and 1200 B.C.E. Beyond the Indian subcontinent, mead achieved widespread popularity in ancient Rome and medieval Europe. In the Anglo-Saxon epic poem *Beowulf,* the Danish queen used elaborate mead ceremonies to bolster political alliances between rival clans. Nine centuries later, the influential anthropologist Claude Lévi-Strauss remarked on the crucial social role mead consumption played among the Caduveo and Bororo tribes in Brazil's Mato Grosso region. To this day, variants of honey mead, such as Ethiopian *tej* and Finnish *sima,* enliven holidays and invigorate celebrations. However, as early as the twelfth century in some parts of Europe, more easily mass-produced competitors—grape wines and grain- or hop-based beers—displaced mead.

The decline of honey was by no means as abrupt. It was not until the emergence of the transatlantic slave trade in the fifteenth century, and the ensuing globalization of sugarcane production, that refined sugar superseded honey as the world's dominant source of concentrated sweetness. Subsequent developments, including the discovery of beetroot sugar in the eighteenth century and the invention of high-fructose corn syrup after the Second World War, further diluted honey's share of the global sweetener business.* Honey still remains a frequently used additive, a common cupboard item, and an ingredient in traditional recipes, but it now occupies less than 1 percent of the worldwide sweetener market. While many honey varieties are tasty and aromatic reminders of the connections among insect pollinators,

---

* Since the mid-1990s, sugarcane and sugar beets have accounted for about 45 percent and 55 percent, respectively, of sugar production in the United States.

flowers, and humans, bees are no longer the world's exclusive custodians of sweetness.

The connotations the word "honey" conveys have been more persistent. European writers were using "honey" as a term of affection since at least 1600, the year English playwright John Marston penned his tragedy *Antonio's Revenge*. Toward the beginning of the play, the Duke of Venice asks his servant, "Canst thou not honey me with fluent speech?" Children's books thrive off such associations. In 1926, A. A. Milne introduced the world to a teddy bear named Winnie-the-Pooh whose infatuation with "hunny" was among his most endearing traits.

The food industry also embraces this metaphor, even if the role played by real honey in manufacturers' products has drastically diminished over time. One of America's best-selling breakfast cereals, Honey Nut Cheerios, is predominantly sweetened with sugar (and contains no nuts, just almond flavoring). Among the ingredients, honey comes in a distant fifth, after whole grain oats, sugar, oat bran, and modified corn starch.

Similarly, another insect-derived product, iron gall ink, influenced human culture from antiquity through the early modern era. This indelible, waterproof substance served as Europe's most important ink for the past two millennia. In response to a chemical secretion from wasp larvae of the Cynipidae family, some oak trees (*Quercus infectoria*) produce Aleppo galls, brittle knobs about the size of walnuts. These galls serve as both the habitat and the food source for the maturing cynipid wasps. In the presence of certain fungi, the galls release tannic acids. Ink makers gathered the galls, dried them, and fermented them, mixing their tannin-rich pigments with iron sulfate, water, and a binder to make highly durable ink. This final ingredient was often gum arabic, the hardened, pale pink sap from acacia trees.

Prior to the age of iron gall ink, ancient Romans and Egyptians had relied on carbon inks, usually a blend of soot and water, augmented with plant pigments. These inks retained their color with age but were susceptible to smudging and could be removed from documents with minimal effort. The technologies of writing began to change in the fifth century. Martianus Capella, a Latin poet from Carthage, a city in what is now Tunisia, authored one of the earliest recipes for iron gall ink in 420 C.E. He referred to his formula as *"gallarum gummeosque commix-*

tio," or a mixture of gall and gum.* By the twelfth century, iron gall ink had completely replaced carbon ink throughout Europe and the Middle East. Many of Western culture's most noble documents, including the earliest complete copy of the Christian New Testament (the Codex Sinaiticus), the Magna Carta, the U.S. Declaration of Independence, Goethe's celebrated drama *Faust,* and Mozart's opera *The Magic Flute* were all written in iron gall ink. Rembrandt and Van Gogh sketched in its rich, velvety hues, and Shakespeare penned his sonnets and plays in this insect pigment. As the raucous Sir Toby Belch declares in *Twelfth Night,* "Let there be gall enough in thy ink."

Unfortunately, iron gall ink becomes unstable after prolonged exposure to humid, oxygen-rich environments. Over time, it corrodes, deteriorating the paper underneath. This regrettable feature poses significant challenges for the conservation of cultural heritage and was one of the factors that contributed to gall ink's decline. Another drawback was that it could not be applied uniformly to the metal type of Europe's fifteenth-century printing presses. The advent of thicker oil-based inks and more stable synthetic pigments and dyes inaugurated a new era in printing.

What does it mean to acknowledge that insects were so central to the histories of sweetness and the printed word? Human revelations have always emanated from intricate webs tethered to the nonhuman world. Yet it is a very modern tendency to forget this interdependence. In the cases of honey and iron gall ink, substances that opened new avenues for our self-expression came from creatures that we so often treat as pests, aliens, and foreign bodies.

Unlike these two insect products, shellac, silk, and cochineal are ancient in origin but did not experience a steady decline with the rise of the modern era. Instead, their brief disappearance in the mid-1900s was followed by a vigorous resurgence in recent decades. The soulful notes of Ella Fitzgerald's shellac recordings, the lustrous threads of Sultan Abdülmecid I's silk carpets, and the radiant crimson hues of Brigadier General O'Hara's cochineal-dyed coat are far more than artifacts of a bygone era. Instead, they endure as animate reminders of an existence filled with insects.

---

* Unless otherwise noted, all translations of sources from other languages are my own.

2

SHELLAC

The sensation of being "in the groove" is the holy grail of jazz. As the renowned drummer Charli Persip described it, "When you get in that groove, you ride right down that groove with no strain and no pain—you can't lay back or go forward. That's why they call it a groove. It's where the beat is, and we're always trying to find that." This expression from the Roaring Twenties is an allusion to an insect secretion. The close fit between a phonograph needle and the grooves in early shellac 78 rpm records determined the quality of the playback. Shellac—a resinous, amber-colored secretion of the tiny scale insect *Kerria lacca*—served as the key ingredient in the first generation of phonographic disks. Odd as it may seem, a gummy substance manufactured by bugs and their human hosts in South Asia was the pioneering medium for the transmission of recorded sound.

The curious story of how a sticky discharge from billions of insect bodies became a vehicle for the globalization of audio culture spans millennia and crosses oceans. Shellac first gained attention through its combustible role in the ancient Indian Sanskrit epic *The Mahabharata* (c. 300–400 C.E.). In this fast-paced saga, filled with internecine feuds and palace intrigue, Duryodhana, the leader of one of two feuding royal factions, attempts to murder his cousins. He traps the Pandavas in a flammable house built of shellac, but they foil his plot and escape to safety through a secret passageway.

The name for Duryodhana's combustible shellac—or simply "lac"—

comes from the Sanskrit word *laksha,* which means "one hundred thousand," a reference to the enormous number of lac bugs that swarm on the branches of certain trees. As with silk and cochineal, host plants are the organic intermediaries that facilitate these human-insect relationships. For thousands of years, peasants in India and other parts of Southeast Asia have raised lac insects on three species of fig and acacia trees: kusum (*Schleichera oleosa*), palas (*Butea monosperma*), and ber (*Ziziphus mauritiana*). The branches of these trees nurture two strains of lac insects, known colloquially as *kusmi* for lac raised on kusum trees and *rangeeni* for lac raised on palas and ber trees.

The arc of a lac bug's life depends on its gender. Male lac insects lose their sucking mouthpieces and antennae during their first weeks of life. As if consumed by wanderlust, they develop legs, grow wings, and fly away in search of mates. In stark contrast, wingless, blind females are sedentary during their six-month existence. They grow to the size of apple seeds and colonize the green twigs of host trees. Sap from the tender shoots nourishes the busy bugs as they secrete brittle tunnels around tree branches. These lac passageways act as defensive canopies, shielding the insect's thousands of tiny, scarlet-colored offspring from predators, ultraviolet rays, and inclement weather.

Women and men collect this valuable product, known as "stick lac," by scraping it from tree branches during two annual harvests. They then crush the stick lac and filter, wash, and dry the pellets. At this point, the substance, now known as "seed lac," is ready to be melted and filtered once again. Workers stretch the heated lac—a gooey substance similar to warm toffee—into rectangular sheets the size of bath towels, which they eventually crack into coin-shaped, shell-like flakes, thus the English name "shellac." Shellac also comes in cookie-sized disks known as "button lac." To manufacture just one pound of this precious substance requires the encrusted secretions of 140,000 lac insects.

Throughout Indian history, lac has been prized by healers and artisans. Practitioners of Ayurveda, one of the world's oldest holistic restorative systems, filtered shellac and blended it with other substances to create tinctures and medicated oils for treating a range of ailments, from arthritis to ulcers. Craftspeople coated elaborate decorative bangles and ceremonial terra-cotta dolls with shellac, and painters waterproofed buildings and varnished furniture with it. Often, these applications received official sanction. In 1590, the Mughal emperor Akbar I, a stickler for detail, decreed the necessary proportion of shellac to certain pigments when coating the doors of public buildings.

During the same decade, Indian shellac emerged onto the world stage. Shellac's promoters praised the unique sheen of shellac polish and the vermilion hues that its unfiltered pigments imparted. In 1596, the Dutch chronicler of East Indian trade routes, John Huyghen van Linschoten, marveled at "desks, Targets, Tables, Cubbordes, Boxes, and a thousand such thinges, that are all covered and wrought with Lac of all colours and fashions; so that it maketh men to wonder at the beautie and brightness of the colour, which is altogether Lac." In the workshops of Europe's artisans, shellac and lac pigments mingled with other insect secretions. A seventeenth-century Venetian dyeing manual lists recipes describing how *"a tingere seta con gomma di lacca"* ("to dye silk with lac gum").

By the 1600s, merchants in Venice, Great Britain, China, and Japan were importing enormous quantities of shellac from India's forests. In 1676, the French explorer and diamond trader Jean-Baptiste Tavernier visited Assam—a state in northeastern India, located south of the eastern Himalayas—and remarked on the many uses of shellac products: "That formed on trees is of a red colour, with it they dye their calicoes and other stuffs, and when they have extracted the red colour they use the lac to lacquer cabinets and other objects of that kind, and to make Spanish wax. A large quantity of it is exported to China and Japan, to be used in the manufacture of cabinets; it is the best lac in the whole of Asia for these purposes."

Another of shellac's virtues was its convenience. With relatively little preparation, it could be applied to many surfaces. A seventeenth-century British treatise on furniture making recommended that craftsmen merely mix one and a half pounds of Indian shellac with a gallon of "spirit" (alcohol) and steep the concoction for twenty-four hours. Taking care not to collapse from the intoxicating fumes, the artisans could then strain this vaporous brew and spread it onto wood, plaster, terra-cotta, or a host of other materials, to waterproof, protect, and brighten them.

In contrast to honey and iron gall ink, shellac's popularity did not fade with the arrival of the modern era. A full two centuries before 78 rpm records revolutionized the music world, shellac was shaping eighteenth-century soundscapes. In the late 1600s and early 1700s, two luthiers from the northern Italian city of Cremona began varnishing their exquisitely crafted stringed instruments with shellac. The violins, violas, and cellos that Antonio Stradivari and Giuseppe Guarneri constructed remain among the world's most coveted for their unrivaled acoustic qualities. Recent scientific analyses of a 1734 Stradivari violin and a 1736 cello of similar vintage reveal shellac as a key component of the instruments' finishes and an important contributor to their unique tonal qualities. A paper-thin layer of this insect secretion improved the resonance of the carved spruce and maple sections of the instruments' bodies. Shellac also protected these fragile creations from exposure to moisture as well as the wear and tear of repeated use. The rich, dark bass tones and the brilliant, high-frequency sounds that these much-sought-after instruments create are, at least in part, the product of insects from the forests of South Asia.

The renowned Cremona luthiers and their fellow European artisans acquired shellac from cosmopolitan dealers whose livelihoods depended on the comings and goings of immense merchant ships, known as East Indiamen. These heavily fortified vessels, which often weighed as much as twelve blue whales and traveled in armed convoys to protect against marauders, sailed the treacherous sea-lanes that linked London to the Indian subcontinent, the East Indies, and China. On the bustling Calcutta docks, stevedores working for the British East India Company loaded thousands of shellac-filled wooden chests aboard East Indiamen bound for harbors like Aden and Port Said. The merchandise then traveled onward to European ports. These ships were like floating bazaars. Their extravagant cargoes included an aromatic miscellany of peppercorns, cloves, nutmeg, cinnamon, indigo dye, cotton, silk, tea, coffee, sugar, shellac, and saltpeter (a key ingredient in gunpowder).

The sudden affluence attained by a handful of British merchants in the colonial maritime trade bred resentment among their compatriots. "Nabob" became a derogatory nickname for those who returned to London from India and purchased political appointments with their newly minted fortunes.* English dramatist Samuel Foote popularized this pejorative in his 1772 satirical play, *The Nabob.* Two centuries later, William Safire, Richard Nixon's speechwriter, revived the epithet to take aim at a different target. In 1970, Nixon's vice president, Spiro Agnew, whose relationship with the press was hostile at best, lambasted journalists as a bunch of "nattering nabobs of negativism."

Long before the arrival of this discordant political moment, a class of Anglo nabobs had built extravagant fortunes by linking the daily lives of South Asian peasants with those of metropolitan craftspeople in Europe. Shellac was crucial to the maintenance and growth of these networks. For centuries, lac had been an economic mainstay of rural India's forest-dwelling communities. In 1908, the British colonial official and botanist Sir George Watt remarked, "Lac enters into the agricultural, commercial, artistic, manufacturing, domestic and sacred feelings and enterprises of the people of India to an extent hardly appreciated by the ordinary observer. The existence of the poorer communities, in the agricultural and forest tracts, is made the more tolerable through the income derived from the collection of the crude article." By 1928, at least 750,000 women and men were raising lac insects throughout India.

Many of these people were Adivasis—literally, "original dwellers"— the descendants of the various peoples of the ancient Indus Valley civilization who had fled to the subcontinent's densely forested hill regions when nomadic tribes from central Asia swept into India thirty-five hundred years ago. Composed of more than two hundred distinct ancestral groups and speaking over one hundred languages, Adivasis have long relied on the subcontinent's forests as a source of food, medicines, and marketable products. Shellac is one of these sustainably harvested goods. Another is *tendu patta*, or leaves from East Indian ebony trees (*Diospyros melanoxylon*). Workers wrap these pliable leaves around tobacco flakes to create the bidi, an inexpensive cigarette that has been a mainstay of popular culture throughout India since the early 1900s. Similarly, Adivasis have customarily harvested mahua flowers, medicinal bark, edible fruits, and oil-rich seeds from

---

* The word is an anglicization of *nawāb*, an Urdu name for Muslim viceroys of the Mughal Empire (1526–1857).

the evocatively named Indian butter tree (*Madhuca longifolia*). Mahua blossoms are the main ingredients in a traditional distilled wine that is both nutritious and replete with spiritual meaning. Shellac has long played a similar role to those of *tendu* leaves and mahua flowers, providing an essential supplement to the incomes of millions of Adivasis.

Oddly, the central place of shellac in Indian history did not enhance Western understandings of its origins. In 1563, the Portuguese doctor Garcia da Orta reported that lac was an excretion from flying ants. Other sixteenth-century Europeans mistook lac for woods that yielded a red dye, known in Burma by the common name *lakka* and in the Malay Peninsula as *laka*. Similarly, the prolific seventeenth-century French scholar Claudius Salmasius concluded that the name "lac" arose from a Greek term for red wood. Even foreign scientists living in India's shellac-producing regions found the resinous substance and its six-legged sources baffling. The British medical officer James Kerr published a description of the lac insect in the 1781 *Philosophical Transactions of the Royal Society of London,* admitting that he was unable to distinguish between the males and females of the species. He claimed that birds transplanted lac bugs to tree branches. In the same journal, only ten years later, the Scottish surgeon and botanist William Roxburgh appeared unaware of Kerr's scholarship and used the wrong scientific name (*Chermes lacca*) when referring to the lac insect.

Shellac experienced a dizzying array of introductions to its far-flung consumers, further perpetuating such confusions. In an 1813 treatise on tinting and staining, the American physician and chemist Edward Bancroft portrayed lac dye as a novelty, even though it was a well-established colonial commodity. Likewise, as late as 1915, the author of a North American industrial chemistry manual felt it necessary to inform his readers, "It has been erroneously stated that lac is the dried exudation of a tree, caused by the sting of the lac insect, and is similar to rosin in its origin. As a matter of fact, it is the secretion of the lac insect, and is a product of the assimilation of the tree sap which the insect feeds upon, just as honey and beeswax are produced by the modification of the nectar of flowers by the bee." Even the word "lacquer" caused bewilderment. Nineteenth- and twentieth-century European artisans used the term when referring to shellac, but they also employed the same word when discussing other varnishes, such as the sap of the Chinese lacquer tree (*Rhus verniciflua*).

In part, this confusion reflected the sluggish pace of communications during the Age of Empire, a period when the tyranny of distance still ruled commerce and limited the movement of information. This was not the "friction-free capitalism" that Bill Gates has prophesied for the Internet era. At a time when news crossed the oceans by sail, "the winds of fortune" could dictate the speed of data transmission.

At a deeper level, these misunderstandings represented the intellectual failures of colonists. "Imperial oversight" is among history's most blistering contradictions. Bureaucrats and traders often overlooked as much as they saw while rushing to extract value from the unfamiliar environments and complex cultures of "benighted foreign lands." The three insect commodities whose histories this book explores—shellac, silk, and cochineal—were exceedingly difficult to manufacture. Their production required an intimate knowledge of insect life cycles, regional weather patterns, potential predators, and best harvesting practices. The making of these coveted products also demanded a sophisticated understanding of the host plants on which these creatures dwelled and dined. These entomological and botanical details were interwoven with generations of accumulated wisdom, often transmitted orally by elders.

Whether or not colonists understood or misapprehended its origins, shellac found its way into countless products. As a furniture varnish or a stiffener that hatmakers used when fashioning domed caps from wool or fur, and as the predecessor to moldable plastics, it became indispensable to the craftspeople of the eighteenth and nineteenth centuries. European and North American artisans spoke of how "no substitute can be found" for "the permanent and beautiful lac" and how "pieces of beautifully polished furniture would be less pleasing to the eye were we deprived of shellac."

In eighteenth-century New England, craftspeople frequently purchased the raw precursor to shellac, called "Indian seed lac," and used it as a final coating for hardwoods. In the words of two of the era's leading furniture makers, shellac protected wood by ensuring that "no damp air, no mouldring worm or corroding time can possibly deface it." However, seed lac was often prohibitively expensive for many. Because merchants imported it from India to Great Britain before shipping it to North America's colonies, shellac was subject to heavy British import duties.

After the Second Continental Congress adopted the Declaration of Independence in 1776, North American traders were no longer at the mercy of a British monopoly on goods from the Far East. U.S. markets began receiving direct shipments of Asian products, primarily from China and India. Because of shellac's new availability and its malleable texture, artisans chose this insect commodity as the main ingredient in the molded photographic cases of the nineteenth century. These resilient containers safeguarded the delicate images produced by the revolutionary new technology.

In 1839, Frenchman Louis-Jacques-Mandé Daguerre divulged the details of a curious invention that would forever alter visual representation. That year, the Romantic painter and printmaker shared his discovery with the distinguished members of the French Académie des Sciences. The photographic pictures he eponymously called "daguerreotypes" required a complex procedure that used a highly polished sheet of silver-plated copper to generate a direct positive image in a large wooden camera box. Once the plate had been treated with iodine vapors to make it sensitive to light, it could be exposed under a lens for several minutes. The daguerreotypist then applied mercury fumes and salts to develop the image. Because the process did not produce a negative, each daguerreotype was unique. To protect the delicate, one-of-a-kind daguerreotype plates—and the next generations of photographs, known as ambrotypes and tintypes—manufacturers fashioned ebony-colored "Union cases" from an amalgam of shellac, wood fibers, and tinting agents. These artful containers featured elaborate recessed motifs that ranged from patriotic eagles clutching bundles of arrows to extravagant scenes of swooning maidens and epic battles from Greek mythology. In the 1850s, photographer Samuel Peck secured a U.S. patent for the Union case, which he produced in partnership with the Scovill Manufacturing Company in Waterbury, Connecticut. In addition to providing portable boxes for the first generation of photographic images, such shellac containers became collectibles in their own right.

At around the same time, an entrepreneurial immigrant named William Zinsser was working around the clock to popularize bleached shellac as a translucent wood varnish. Zinsser, who had served as a foreman in a German shellac factory that imported and processed the Indian raw material, immigrated to the United States during the

European revolutions of 1848. At his workshop in Manhattan, Zinsser invented a new way to chemically remove shellac's pigments. He promoted his new product under the refined name "White French Varnish." Zinsser initially marketed his quick-drying gloss to his fellow German-speaking furniture makers and carpenters in New York's burgeoning immigrant districts. The novel creation soon became popular with a wider range of customers throughout the United States. With nothing more than a paintbrush and a drop cloth, an amateur restorer working out of a basement or a garage could apply Zinsser's varnish and instantly add a striking patina to fine-grained woods.

Shellac entered the living rooms and lexicons of many more consumers after 1896, when Emile Berliner discovered its utility to the recording industry. Like Zinsser, Berliner was a German-born inventor who had come to the United States at midcentury. Berliner found that warm shellac could be molded and machine-pressed into hard disks for radio and home-audio playback. Shellac is a natural thermoplastic, pliable when heated and solid at room temperature. In addi-

A shellac "Union case" from the 1860s, showcasing one of the many
uses of the hard but malleable secretions from the lac insect (*Kerria lacca*).
Such cases protected fragile, one-of-a-kind daguerreotype images
and were collectors' items in their own right.

tion, it mixes well with fillers and dyes. As a result of its favorable physical properties and its relative abundance, shellac offered a promising substitute for the materials used by the nascent recording industry. In the 1870s, Thomas Edison had developed the first phonographic system, which relied on thick brown cylinders of mineral wax. "Edison Gold Molded Records," as they were known, delivered sounds to a flaring horn by vibrating a stylus along the finely etched grooves that ran around the outer surface of the canister. These wax drums were bulky, expensive, and difficult to store. Despite a flurry of advertisements that touted "the sweet tone for which the Edison [phonograph] is famous," inventors searched for an alternative material for transmitting musical performances into the parlors of the public.

Berliner patented the gramophone in 1887, convinced that his comparatively light shellac-based platters would replace Edison's cumbersome wax cylinders. Phonograph manufacturers agreed and soon began pressing thousands of recordings into ten- and twelve-inch disks. Rotational speeds varied among manufacturers, but most turntables spun at a brisk 75 to 80 revolutions per minute (rpm). Eventually, 78 rpm became the industry standard.

By 1900, shellac records had eclipsed wax cylinders (and the short-lived vulcanized rubber disk) as the medium of choice for commercial music reproduction. Shellac was the binding agent in a composite that also included powdered limestone, lubricants, and abrasives, to keep the phonograph needle from slipping. The early 78 rpm records featured coarse grooves that generated a crackling surface noise like the sizzle and pop of frying bacon. Record production drove increasing demand for shellac. In 1920 alone, the United States imported more than $23 million worth of shellac (11,568 tons) from Southeast Asia.

Such enormous quantities of shellac originated from vast stands of fig and acacia trees and their insect colonies, which had been carefully tended by villagers across northern India, Thailand, Burma, and the Malay Peninsula. Prior to the 1930s, rural cultivators relied on traditional methods for transforming harvested lac into high-grade shellac. After workers ground, sifted, and washed the stick lac by hand, the resulting piles of amber-colored chips were ready for melting. In 1924, the American actress and travel writer Elizabeth Brownell Crandall vividly described the ensuing process: "The mixture is then put

A Columbia shellac 78 rpm record from 1925. During the first half
of the twentieth century, a resinous substance produced by insects and their
human cultivators in South and Southeast Asia was the key component
in the world's first widely circulated medium for recorded sound.

into long, narrow cloth bags, ten to twelve feet long and two inches
wide. These wormlike bags are held over open charcoal fires by two
operators who begin to twist the bags in opposite directions, while
the melted Lac slowly oozes out and drops upon the floor." Crandall
continued: "As the melted Lac drops to the floor, it is spread out by
means of a pineapple leaf. Before it has time to congeal it is picked up
by still another native, who stretches it into thin sheets, placing a foot
on either end of the piece and then pulling it upward by means of his
teeth and hands." The process Crandall described was typical of shellac
production in the 1920s.

Although aspects of this grueling procedure became mechanized
as the twentieth century wore on, many shellac exporters continue to
maintain that handmade shellac is of a much higher quality than its
machine-made counterpart. To this day, workers at thousands of local
shellac operations in India still melt lac over charcoal-burning clay
ovens known as *bhattas,* and they continue to hand-stretch the molten
drippings that seep from the long, rotating cotton bags.

From these *bhatta* furnaces flowed the raw material that record companies pressed into the first generation of shellac records. The pioneering artists of the 78 rpm era were a diverse group. Their musical range spanned the big, brassy sounds of the young Louis Armstrong and King Oliver's Creole Jazz Band in the 1920s to the stirring renditions of Jean Sibelius's symphonies that the Finnish symphonist Robert Kajanus conducted during the 1930s. Such sonic milestones marked the onset of a new era, one that gave an afterlife to the formerly unrepeatable performance. In the words of Bruno Walter, one of the twentieth century's greatest conductors, "Recording is the only form of immortality attainable by a performing musician." With this transformation in the reproduction of sound, music experienced a revolution.

Although the shellac 78 extended countless new soundscapes to artists and listeners, it also limited musicianship in unforeseen ways. Igor Stravinsky's 1925 *Serenade in A for Piano* exemplifies this paradox. The Russian-born composer tailored each of the four movements of his *Sérénade* to conform to the three-minute time constraints of a ten-inch 78 rpm shellac record. Stravinsky recalled, "In America I had arranged with a gramophone firm to make records of some of my music. This suggested the idea that I should compose something whose length should be determined by the capacity of the record. I should in that way avoid all the trouble of cutting and adapting. And that is how my *Sérénade en la pour Piano* came to be written."*

Likewise, jazz musicians of the 1920s and 1930s pruned their sprawling dance hall tunes to fit on the short sides of a shellac record. Legendary New Orleans drummer Warren "Baby" Dodds was hardly alone in resenting these restrictions on his creative freedom. Recounting the character of the recordings he made with his brother Johnny's band, Dodds remarked, "We could never play as many choruses as we used in dances, and if there were solos they had to fit into the exact time, too." Ironically, Dodds was more "in the groove" when performing live than when his beats were being etched into the deep furrows of a shellac record.

---

* The "LA" in the title of Stravinsky's *Sérénade* is one of the syllables from solfège, a technique in Western music where each scale degree is assigned a coordinating syllable. The most famous solfège is: do, re, mi, fa, so, la, ti.

Shellac 78s faced other challenges. Because these records had such brittle surfaces, they exhibited an unusual degree of embedded obsolescence. The tracking device on a typical Victor phonograph put fifty thousand pounds per square inch of downward pressure on the steel needle, raking the grooves of the shellac disk like a farmer's plow carving through clods of tender soil. Cosimo Matassa, founder of the prolific J&M Recording Studio in New Orleans, recalled that in the 1940s, "A hit record would get worn out each week on the jukebox, because shellac supplies were scarce and record formulations were poor. . . . [A record] got played 100/110 times and it was worn out. So a hit record kept on selling and selling." For music companies, this was quite a windfall. Consumers could hardly avoid coming back for more. It also kept many insect communities and their human hosts in business.

As serial record buyers were often surprised to learn, their audial habits depended on the bodies of bugs. Referring to shellac's insect producers as "Mr. and Mrs. Lacca," the author of a 1937 article in *Popular Mechanics* explained that these creatures "still retain the world monopoly on the shellac business. When man needs real shellac today, he still must rely on a bug—and like it." However, such six-legged industries were not immune to geopolitical upheaval. During the Second World War, shellac supplies underwent a drastic contraction. German U-boat attacks on Allied merchant ships and Japan's invasions of the Malay Peninsula, Thailand, Indochina, and Burma disrupted global supplies of the valuable product. The United States War Production Board began rationing shellac in April 1942. The substance was a key waterproofing material for the insulated wires and wooden components of midcentury boats and airplanes.

Shortages propelled a transatlantic wartime recycling movement that prefigured the first Earth Day by nearly three decades. In a 1943 campaign dubbed "New Records Depend on You!" a consortium of recording companies urged owners of unwanted 78s to return their shellac disks for reprocessing. In Britain's *Gramophone Magazine*, representatives of Decca Records, Columbia, and Parlophone announced, "Owing to war conditions the Government has found it necessary to conserve supplies of shellac and other materials essential for manufacturing records by the most stringent restrictions as to the use of these materials." The statement continued with an exhortation: "The further maintenance of adequate record supplies will depend upon the good-will and readiness of the public to return old and unwanted records, because only by this means will manufacture continue." By May 1946, U.S. shellac prices had hit $45 per ton, more than tripling their prewar average of $14.

After the war, the rise of vinyl heralded the dawn of a new era in recording technology and the conclusion of shellac's half-century reign over the record industry. In 1948, Columbia Records introduced the "Vinylite" 45 rpm, seven-inch extended play ("EP"). Shortly thereafter, RCA Victor followed suit with the 33⅓ rpm, twelve-inch long-playing ("LP") vinyl. New records, created by synthesizing a copolymer of vinyl chloride and vinyl acetate, were harder and smoother than those made of shellac, allowing manufacturers to press more grooves into the disk. Shellac 78s contained around 85 grooves per inch, while the newer EPs and LPs averaged between 224 and 260 grooves per inch. This innovation transformed the listening experience by extending playing time, reducing background noise, and enhancing record durability. The "hi-fidelity" era had begun.

Even so, the shift from shellac to vinyl was not an overnight phenomenon. As late as 1953, *The New York Times* reported that U.S. sales of shellac 78s had totaled $89.7 million the previous year, exceeding the combined sales of EPs and LPs by more than $6 million. This was, however, the last year the shellac 78 outperformed its vinyl competitors in the United States. Commercial production of shellac 78 rpm records in the United States ended in 1958.

In retrospect, the seventy-year-long heyday of the shellac disk was an era of unprecedented longevity for a musical technology. From the advent of shellac records in the 1890s to their ultimate demise in 1962—the year when EMI Music withdrew its last remaining 78s from its catalogs—these brittle grooved platters filled the shelves of music listeners worldwide. The sonic technologies that have followed—vinyl records, reel-to-reel audio, eight-track recordings, cassette tapes, compact discs, DATs, and MP3s—have all had much shorter commercial lives.

The demise of the 78 rpm record did not curtail shellac's global reach. It experienced a full-blown revival after the Second World War. In fact, shellac's presence in the vernacular of U.S. politics never waned. One of these expressions was on vivid display during the 2010 midterm elections, when the Democratic Party suffered demoralizing losses in congressional and governors' races throughout the United States. Speaking to the press corps in the aftermath of the defeat, President Barack Obama accepted full responsibility for his party's electoral woes: "I'm not recommending for every future President that they take a shellacking . . . like I did last night." Obama employed a colloquialism that dates to at least the 1920s. In all likelihood, shellac's widespread association with the stupor-inducing alcohols that finished a project (or vanquished the craftsman's clearheadedness in the process) prompted North American sportswriters to use the term when describing boxing knockouts and baseball blowouts. The earliest instance of this buzzword for defeat is a June 25, 1923, headline from Connecticut's *Hartford Courant* that reads, "Luque's Streak Ends When Cubs Shellac Reds, 2 to 0." The English language is capable of remarkable contortions. Almost simultaneously in the 1920s, the insect-inspired expressions of "being in the groove" and "getting shellacked" emerged at opposite extremes of the pleasure-pain spectrum.

"Jukebox" was another evocative, shellac-related term coined during the decade of the Harlem Renaissance, flappers, and the speakeasy. The word "juke"—insinuating disorderly or wicked things—comes from Gullah, a creole of English and West African dialects long spoken by descendants of African slaves in coastal South Carolina, Georgia, and northeastern Florida. While in the nineteenth century, "juke house" or "juke joint" initially referred to a roadhouse or brothel, the term "jukebox" eventually came to mean a public gramophone. The first of these so-called "nickel-in-the-slot" machines (not yet called jukeboxes) debuted on November 23, 1889, at the Palais Royale Saloon in San Francisco. Manufactured by the Pacific Phonograph Company, it consisted of an oak cabinet that housed an Edison Class M electric phonograph. After depositing a nickel, a listener could hear music played from the single wax cylinder inside the device. Since the mechanism lacked amplification, stethoscope-like rubber tubes functioned as crude headphones. The device looked like a fanciful creature from a Jules Verne adventure.

By the late 1920s, a new version of this coin-operated gramophone record player, now filled with towering stacks of shellac 78s, was enlivening roadside diners and animating dance halls across the United States. The jukebox also found a receptive home in that quintessentially American cultural institution, the bowling alley. In fact, during the first half of the twentieth century, the nation's ten-pin alleys literally oozed shellac. In addition to the shellac records spinning on turntables, the bug secretion coated the wooden lanes, which needed frequent applications of clear varnish to keep the balls rolling smoothly. As one longtime arcade worker from Wisconsin recalled, "After shellacking the alleys at night we had to be careful to turn out the lights so the heat generated by the lights wouldn't start a fire." It is unlikely that Duryodhana's flammable lac house came to mind when bowling alley owners envisioned such catastrophic scenarios.

In the 1940s, polished nails came into vogue. The American actress Rita Hayworth's glamorous fashion sensibilities deserve the lion's share of the credit for this development. Her trademark carmine lips and matching red nails, which appeared in full Technicolor glory in films such as *Blood and Sand* (1941), worked wonders for the cosmetics industry. Fast-drying shellac was the main ingredient in the crimson-hued manicure glosses of the day. It was also the primary component of the first commercially available aerosol hair sprays. Although they were likely oblivious of it at the time, Americans were coating their hair and nails in an insect secretion, grooving to its resonant tones, and bowling strikes and spares on its smooth veneers.

Shellac's heyday was cut short by the arrival of a new generation of man-made substitutes during the postwar years. From the 1950s onward, vinyl records, urethane wood varnish, acrylic nails, and an array of other synthesized polymers dethroned the insect secretion from its position as the world's leading proto-plastic.

However, just as shellac seemed to have reached its coda, this ancient insect secretion once again came to occupy a central place in the daily lives of consumers around the globe. During the late twentieth century, discoveries showing that many of the synthetic substitutes for this insect secretion were toxic to the human body and harmful to the environment opened a door for the reemergence of shellac in a vast array of venues. A stroll through the aisles of any North American pharmacy, hobby shop, supermarket, or convenience store makes this abundantly clear. In pharmaceutical products, shellac serves as an enteric coating to slow the digestion of medicines in the acidic environment of the human stomach. Shellac is sold as a popular varnish for furniture and decks, it keeps the skins of citrus fruits and apples waterproof and shiny, it adds a glossy patina to candies, and it augments the drying properties of many types of nail polish, hair sprays, eyeliners, and mascaras. It appears on the ingredients lists of dentures and tooth fillings, and it is increasingly often used as a nontoxic preservative for cadavers.

Shellac is, quite literally, everywhere, from our hair and teeth to our fingernails and stomachs (even after death). Whether we are aware of it or not, we are all "in the groove," moving to the rhythms of the lac bug's life cycle.

3

SILK

Operagoers the world over recognize the allure of Puccini's Cio-Cio San, or Madama Butterfly. "Light as a feather she flutters, / And, like a butterfly, hovers and settles." Less familiar is the aesthetic ideal of Madame Moth.* The feathery, scent-gathering antennae of an adult silkworm moth might seem alien, even creepy, to contemporary Westerners. Yet in other times and places, these curvy insect feelers have epitomized human desire. A venerable Japanese proverb declares, "The silkworm-moth eyebrow of a woman is the axe that cuts down the wisdom of man." Likewise, the oldest surviving collection of Chinese poetry, *The Book of Odes,* features an entomological tribute to a noblewoman's face: "Her forehead cicada-like / her eyebrows like [the antennae of] the silkworm moth / What dimples, as she artfully smiled!" Throughout China's imperial history, the attraction of a woman's *éméi* (蛾眉)—her "moth-feeler eyebrows"—was a persistent theme. The moth's crescent-shaped antennae even inspired the naming of the arching peaks atop one of China's four sacred Buddhist Mountains, Mount Emei in Sichuan Province. Such fashionable associations with silkworm eyebrows may strike modern-day readers as eccentric.

* Moths and butterflies both belong to the order Lepidoptera. One way to tell the difference between them is that a moth's antennae are feathery or saw-edged, while a butterfly's antennae are club-shaped, with a long shaft and a bulb at the end. Another of their distinguishing features is that moths tend to hold their wings in a tentlike fashion that hides their abdomen; butterflies generally fold their wings vertically up over their back.

A fully grown silkworm (*Bombyx mori*) moth.

Much as standards of beauty have waxed and waned, so have the fortunes of insects in human history.

The silkworm (*Bombyx mori*) embodies a more complete domestication than any of the other animals in the typical barnyard menagerie. The "worm" in its name is a misnomer. It is really the larval stage of a caterpillar, entirely dependent on humans for its survival. Like most insects, silkworms undergo four developmental phases: egg, larva, pupa, and metamorphosis into an adult. Following this final transformation, the blind and flightless cream-colored moth that emerges from its woolly cocoon is the product of so much selective breeding by humans over thousands of years that the insect cannot feed itself. Procreation on an empty stomach serves as the coda of its six-week life. Like bar patrons at last call, male moths flutter about clumsily, intoxicated by the pheromone signals emanating from eligible mates.

Most silkworms never achieve reproductive climax. Instead, their lives come to an inglorious end in a cauldron while, as pupae, they are still swaddled in their cocoons. Sericulturists—to use the technical term for people who harvest and process the silk—extract the coveted fibers by "reeling" them from pupae soaking in a bath of boiling water. This step removes the gum that binds the cocoon's filaments, ends the silkworm's existence, and begins the journey that its cocoon thread will make into an elaborate web of human producers and consumers.

Complex relationships that transcend species boundaries are always life-changing for both parties. This remains especially true with the enduring bond between silkworms and humans. Many civilizations have tailored their habits, traditions, and sacred rituals to the needs of these fastidious creatures. The word for such border crossings, "domestication," suggests an invitation into the *domus*, Latin for "home." To that end, silkworms have crossed the threshold, sharing beds with their human companions, occupying choice spots by the hearth, and entering countless houses of worship. In the 1850s, the Irish travel writer Mabel Sharman Crawford recounted how Tuscan women would tuck silkworm eggs into their blouses and carry them to church on Sundays to keep these future thread spinners warm. Half a world away, nineteenth-century Mormons in the American West did the same thing. One such colonist, Priscilla Jacobs, attended worship services at her tabernacle in northern Utah with her unhatched silkworms stored in a bag around her neck. Jacobs was startled when the larvae began to emerge during the sacrament. She excused herself and hurried home to care for her squirming charges.

These Italian Catholics and North American Mormons were following long-standing precedents inherited from the Chinese peasants who first developed techniques of sericulture, nearly five thousand years ago. Chinese families who reared silkworms on shallow, elliptical trays woven from bamboo strips coddled their larvae like newborn infants. This attentiveness is on vivid display in the canonical rules for silkworm care, handed down over countless generations:

> The eggs when on paper must be kept cool; after having been hatched they require to be kept warm; during their period of moulting they must be kept hungry; in the intervals between their sleeps they must be well supplied with food; they should be kept dark and warm; after they have cast their skins, cooled and allowed plenty of light; for a little time after moulting they should be sparsely fed and when they are full grown ought never to be without food; their eggs should be laid close together, but not heaped upon each other.

On cold nights, Chinese peasant families often surrendered their bedrooms to their insect companions, even sleeping in the barn so that their precious caterpillars could snooze in warmth and comfort. Women and men living in close proximity to silkworms forswore tobacco and garlic for fear that strong odors would disturb the larval growth and cocoon production. Ancient Chinese silkworm manuals even recommend tickling drowsy, newly hatched larvae with chicken feathers to stimulate their development.

The tapestry of human-insect convergence is woven in long, shimmering threads. From worm spit comes unparalleled beauty. A mature silkworm's two salivary glands secrete a pair of continuous filaments (called "brins") that stretch up to three thousand feet, the distance of ten U.S. football fields. These brins are composed of an exceptionally stretchy protein called "fibroin," surrounded by a gummy compound known as sericin. A distinctive blend of patience and precision sustains the silkworm's fabrication of the cocoon from these sticky fibers. The three-inch-long silkworm gyrates its head in a figure-eight motion for three or four days, releasing its thread-forming gel at an astounding rate of four to six inches per minute. This process yields as much as thirty feet of fiber every hour. After fabricating a stretchy "hammock" of anchor threads to affix its body to the nearest stable surfaces, the silkworm constructs its protective scaffolding by crisscrossing its filament in matted layers. As the gluey sericin is exposed to air, it hardens to the consistency of Styrofoam. This lozenge-shaped bundle will shelter the tawny pupa during its three-week metamorphosis into a full-grown moth.

Unless it is one of the few selected for mating by its human handlers, the silkworm's metamorphic ambitions end here. Sericulturists immerse fresh cocoons in hot water to kill the pupae, rinse the sericin, and unwind the fibroin strands.* Following these steps, they "throw" the long silk fibers, intertwining from two to as many as ten strands to thicken the final thread for weaving. Because silk threads feature a

---

* In some cultures, the pupae become food, while in others they are composted or thrown away. In 2002, a government officer from India's Andhra Pradesh state developed ahimsa silk, or peace silk. This method of silk production allows the caterpillar to hatch before a worker unreels the cocoon thread.

A statue of Empress Leizu, the legendary "discoverer of silk," at the Suzhou Silk Museum in eastern China's Jiangsu Province. She is holding a mound of cocoons in her right hand. Exhibits showcase the five-thousand-year relationship between silkworms and humans in Chinese history.

molecular arrangement of tightly packed proteins, the resulting filaments are supple yet exceedingly resilient. Under the right conditions, they can achieve the tensile strength of steel and the pliability of rubber. In cross section, silk strands have a triangular prismlike structure. This unusual profile causes silk fabric to refract incoming light at different angles, producing stunning visual effects. Unlike its coarser, short-fiber relatives in the fabric world—cotton, linen, and wool—silk is smooth and light, featuring long, continuous strands. In these regards it is also superior to other luxury animal fibers such as alpaca, mohair, and vicuña. The extraordinary properties of silk have heightened its allure. For thousands of years, a secretion from the salivary glands of tiny caterpillars has driven government policies, inspired poetic odes, and galvanized imperial ambitions.

Indeed, it is an empress who often receives credit for discovering silk's unique attributes. According to Confucius, in 2640 B.C.E., Emperor Huangdi's wife, Leizu, was steeping her afternoon tea when a silkworm cocoon inadvertently tumbled into her cup from the branch of an overhanging mulberry tree. Leizu tugged on the loose end of a thread, unraveling a malleable fiber that stretched the length of her garden. Recognizing that mulberry leaves were the caterpillars' preferred food, Leizu planted a grove of white mulberry (*Morus alba*) trees and began cultivating silkworms for their precious filaments. She also invented a loom for weaving these delicate threads. Thus, China's silkworm tradition was born.

This legend has undergone many incarnations. In his 2002 novel, *Middlesex,* Jeffrey Eugenides traced an imaginative (and imaginary) four-thousand-year historical arc from Leizu's tumbling silkworm cocoon to Isaac Newton's falling apple. "Either way," he wrote, "the meanings are the same: great discoveries, whether of silk or of gravity, are always windfalls. They happen to people loafing under trees." Despite this lyrical symmetry, the apple and the "worm" pull in opposing directions; ripe fruit falls to the ground, but silk gracefully defies gravity's compulsions. Its metaphorical and material buoyancy is on vivid display in "Silk Parachute," John McPhee's meditation on his relationship to his mother and the passing of his youth. The story ends with a touching description of a favorite childhood toy. When thrown, this rubber ball split into two hemispheres, releasing a small silk canopy: "Folded just so, the parachute never failed. Always, it floated back to you—silkily, beautifully—to start over and float back again. Even if you abused it, whacked it really hard—gracefully, lightly, it floated back to you."

Archaeologists can now confirm that the weightlessness and grace of silk stirred the human spirit for at least three millennia prior to Leizu's teacup epiphany. Approximately 5,650 years ago in east-central China's Yellow River Valley, a community known as the Yangshao buried a child wrapped in a silk shroud. The funerary garment, woven with gossamer-thin filaments painstakingly gathered from wild silkworm (*Bombyx mandarina*) cocoons, must have been extraordinarily valuable to its ancient creators. Even in an era where the specter of death lurked around every corner, anguish accompanied the loss of a youngster. Given the hardscrabble existence of China's Neolithic farmers, material tributes were in short supply. A precious object like handwoven wild silk provided a poignant marker of an untimely death.

Silk's value only increased as its reputation spread. We are unaccustomed to thinking of insects and their secretions as money, but silk—like beeswax—experienced a protracted heyday as a medium of exchange. This was the case in the elaborate extortion rackets typical of China's early imperial history. During the Han Dynasty (206 B.C.E.–220 C.E.), government envoys sent innumerable bolts of the shiny fabric to the northwestern reaches of the empire to dissuade their foes from attacking vulnerable Han border towns. The beneficiaries of these lavish bribes were the Xiongnu, a confederation of pastoral nomads who inhabited the vast grasslands of the East Asian Steppe. The efficacy of the Han Dynasty's silk-lined diplomacy was a matter of considerable debate. In the reign of Emperor Wen (202–157 B.C.E.), the young Han poet-scholar Jia Yi grumbled, "Now the [Xiongnu] are arrogant and insolent on the one hand, and invade and plunder us on the other hand, which must be considered an act of extreme disrespect toward us. And the harm they are doing to the empire is extremely boundless. Yet each year Han provides them with money, [raw] silk floss, and fabrics." Others maintained that this tribute strategy was a strategic measure for holding adversaries at bay until the Han could build a military capable of confronting the Xiongnu on the battlefield.

Elsewhere, silk also made a lasting impact as imperial armies clashed along contested frontiers. The Romans first encountered the glistening insect secretion in 53 B.C.E. when General Marcus Licinius Crassus and his seven Roman legions suffered a devastating defeat at Carrhae—now known as Harran, in southeastern Turkey—by soldiers of the Parthian Empire, a major political and cultural power that stretched from Bactria (present-day Afghanistan) in the east to the Euphrates River (in what is now Iraq) in the west. Crassus, Rome's wealthiest man, had sought to enhance his prestige on the battlefield. Poor planning and a severe misjudgment of his adversary's strength thwarted his ambitions. In the blistering desert heat, the Roman troops were petrified when the Parthian standard-bearers unfurled their gleaming scarlet silk banners, embroidered with gold threads. This crimson spectacle, accompanied by the deafening thunder of Parthian kettledrums, was probably the first time most Romans had ever seen the glistening fabric. Romans also encountered the Persian compound bow for the first time that day. Plutarch offered a distressing account of the "Parthian shot," with which a hail of arrows from ten thousand mounted Parthian archers pinned many terrified legionnaires to their shields. Incidentally, Carrhae is the birthplace of a scathing retort. When we "take a parting shot" at someone, we are unwittingly emulating Parthian cavalrymen firing on their stunned Roman opponents in 53 B.C.E.

This harrowing first encounter did not deter the ancient Romans from obsessing over this exotic textile from the East. Over time, silk's popularity only grew. The third-century emperor Varius Avitus Bassianus, posthumously known as Elagabalus, refused to wear linen or wool, believing that they were the lowly fabrics of beggars and ruffians. Instead, the devout sun worshipper and renowned transgressor of gender norms appeared in revealing robes of pure Chinese silk. Elagabalus so zealously adhered to Syrian prophecies of his death by violent uprising that he "prepared cords entwined with purple and scarlet silk, in order that, if the need arose, he could put an end to his life by the noose." He did not meet his fate at the end of a caterpillar's thread, however. Instead, Elagabalus died at the hands of soldiers of his own Praetorian Guard, who resented his eccentricities and were incensed by his denigration of Roman customs.

Initially, luxurious Chinese fabrics were reserved for Rome's wealthy, but they soon found favor with a wider public. By the fourth century, one observer of the Byzantine Empire reported, "The use of silk which was once confined to the nobility has now spread to all classes without distinction, even to the lowest." However, while Romans spent astronomical sums on Chinese silk, their knowledge of sericulture remained rudimentary. In his encyclopedic *Natural History,* Pliny the Elder referred to the Chinese as the Seres, or "silk people," who were "so famous for the wool that is found in their forests. After steeping it in water they comb off a white down that adheres to the leaves. . . . So manifold is the labour employed, and so distant are the regions which are thus ransacked to supply a dress through which our ladies may in public display their charms." In all likelihood, Rome's leading natural philosopher had confused silk with Indian cotton, another sought-after fabric from a distant land.

The commodities Romans obtained from these remote kingdoms were often the product of arduous and anonymous labors. A calculus beyond the limits of human understanding would be needed to tally the myriad fingers that have tugged and reeled in these diaphanous threads over the past five thousand years. Conjecture aside, we know with certainty that rural women, most of them Chinese peasants, were the heart and soul of textile craftwork. The Confucian idiom "Men plough, women weave" (*Nangeng nüzhi*, 男耕女織), conveyed the gendered division of labor that governed silk production for much of China's imperial history.

A twelfth-century Chinese poem by a Song Dynasty magistrate named Lou Shou evocatively portrayed the connection between women's labor and the stages of silk thread production. While the stanzas glamorize these duties, they also enrich our sense of the aromas, sounds, and sensations of this rural enterprise:

Throughout the village, the sweet fragrance of cooking cocoons;
Which families' girls have taken on this task?
The charm of the brimming pot fills the kitchen;
Pat, pat, hands check the boiling water.

From above the pot, the color looks right;
Turn the roller, how the yarn is long.
Around evening, they get a short rest;
The chatter of the working girls floats over neighbors' walls.

The insects themselves also wrought the soundscapes of silk production. Caterpillars seem unobtrusive. Even so, a tray of plump silkworms dining on a heap of mulberry leaves sounds like thousands of Rice Krispies popping in a giant bowl of milk. When preparing to build its protective canopy, a silkworm eats as often as ten times a day. Its body weight will increase ten thousand times over the course of its four- to six-week life. By comparison, such growth for a human baby with a birth weight of seven pounds would produce a seventy-thousand-pound adult!

Silkworms are not only voracious, they are notoriously finicky eaters. In fact, their distinctive dining habits are reflected in their scientific name. *Bombyx mori* is Latin for "silkworm of the mulberry tree." The story of silk is also a tale of trees. Arboreal landscapes around the globe testify to this legacy. Aspiring silk makers the world over have committed much time and energy to planting and nurturing white mulberry saplings. In Argentina, Afghanistan, southern India, southern France, and elsewhere, the remains of these groves are ligneous traces of various attempts—some successful, others less so—to raise domesticated silkworms.

In 1901, a noted British explorer of central Asia, Sir Aurel Stein, unearthed the gnarled ruins of several sprawling mulberry groves in the arid Tarim Basin of Xinjiang, in what is now northwestern China. His discovery added another chapter to the story of how mulberry trees, their silkworm consumers, and their human caretakers spread around the world. In this case, a tantalizing blend of myth and fact suggests that silkworm smuggling was an indispensable enterprise in antiquity. Despite the atmosphere of secrecy that shrouded silk production techniques from the outside world, the Chinese empire was unable to retain a permanent monopoly on this knowledge. The oasis kingdom of Khotan, a vital trading hub situated between the bone-dry Taklamakan Desert to the north and the desolate Kunlun Mountains to the south, was the first realm to benefit from the illicit transfer of silk caterpillars. As with Leizu's teacup discovery, a female protagonist takes center stage in this story.

Sometime during the first century c.e., a Chinese princess who later took the name Punesvara married Khotan's Buddhist king Vijaya Jaya. The king sent an envoy to China to escort the bride-to-be back to his palace. The emissary informed Punesvara that Khotan was unable to produce the same quality of fabrics to which she had become so accustomed in her homeland. Upon her departure, the noblewoman violated imperial prohibitions against conveying the materials and techniques of sericulture outside the Chinese empire. In a daring act of subterfuge, Punesvara folded several mulberry seeds and silkworm eggs into the lining of her headdress (likely made of silk), which safeguarded her biotic contraband from the prying eyes of the frontier guards. Once in Khotan, she donated her precious cargo to her adopted kingdom, thereby extending the practices of silk cultivation, harvesting, and manufacturing to the outside world. As a Chinese Buddhist pilgrim later recounted, "The secret of making silk was taken from China, and the people of Khotan began to wear fine silk clothes along with their furs."

In subsequent centuries, other empires also employed trickery to acquire silk-making techniques and technologies. The Byzantine scholar Procopius related how, in 552 C.E., the Eastern Roman emperor Justinian persuaded two Christian monks (most likely members of the Nestorian Church) to smuggle silkworm eggs and mulberry seeds from China to Byzantium in the hollow interiors of their bamboo walking sticks, "and in the manner described caused them to be transformed into worms, which they fed on the leaves of the mulberry; and thus they made possible from that time forth the production of silk in the land of the Romans." The difficulty of transporting silkworm eggs in hot environments sheds doubt on the credibility of this story. Regardless of whether the canes of these traveling brethren actually nurtured Byzantine sericulture, Constantinople's silk industry began in the sixth century and thrived for more than half a millennium.

The closely guarded secrets of silk manufacturing took even longer to reach lands farther west. During the Second Crusade (1147–49), King Roger II of Sicily mercilessly sacked the Byzantine cities of Corinth and Thebes, captured the resident silk weavers, and seized their horizontal looms.* Roger showed little clemency in these invasions, extracting every last ounce of wealth and demanding total subservience from the residents of conquered lands. Despite his notoriously brutal tactics, Roger was a multilingual cosmopolitan who had been tutored by Greek and Muslim instructors. The Sicilian king was well aware of silk's international renown and wasted no time in establishing his own textile operations in Palermo and Calabria.

The forced migration of weaving techniques to Italy did little to disrupt Eurasia's vibrant silk trade. Chinese silk attained such unmatched quality that Europeans could do little but search endlessly for commodities to trade for the prized fabric. The nineteenth-century German geologist Baron Ferdinand von Richthofen coined the term *die*

---

* This type of loom, invented during the eleventh century, allowed weavers to produce longer and wider fabrics at faster rates. It also facilitated the production of more complex patterns.

Seidenstraßen, or the Silk Roads, to name the most enduring commercial corridors in world history. This expansive web of trade routes, spanning more than five thousand miles from Xi'an in northwest China to the shores of the Mediterranean, flourished from 100 B.C.E. well into the fifteenth century.

Surprisingly, this intricate network of camel trails, trading outposts, and frontier towns was maintained by the Mongols. Genghis Khan (1162–1227) may not be the first historical figure who comes to mind when we think of peace and stability, yet his conquests in the early 1200s produced a lasting era of administrative dependability in central Asia. The *Pax Mongolica,* which followed the consolidation of the Mongol Empire, ensured a relatively unhindered flow of commerce among Europe, central Asia, and China. In the romantic language of one contemporary Persian traveler, "a maiden bearing a nugget of gold on her head could wander safely throughout the realm." By 1257, more than a decade before the renowned Venetian explorer Marco Polo visited Cathay (as many European travelers called northern China), Chinese raw silk had already arrived in Tuscany.

Silk also traveled eastward across the Pacific Ocean. Beginning in the late 1500s, a type of silk dress known as the *china poblana* gained widespread popularity in the Viceroyalty of New Spain, the Spanish colonial territories north of the Isthmus of Panama. Seamstresses from the Mexican state of Puebla crafted these ornate women's dresses from Chinese silk. These popular garments, often paired with a white blouse and a silk shawl, became Mexico's national costume in the first decades of the twentieth century.

Sixteenth-century Mexican dressmakers obtained the coveted textile meticulously produced by peasants in rural China thanks to the Acapulco-bound Manila galleons, at least one of which sailed annually for two and a half centuries (the last one departed in 1815). Following Spain's successful conquest of Manila in 1571, the Pacific served as an aquatic highway for lucrative trade voyages. Silver from Spanish colonial mines in Potosí (in modern-day Bolivia) and Zacatecas, Mexico, traveled west on the galleons. Once these lumbering cargo ships had docked at Manila, Chinese merchants exchanged jade, tea, porcelain, lacquered goods, and silk for precious chests filled with South American silver coins. This transaction satisfied the European demand for exotic Asian luxury goods and dramatically enlarged the silver reserves of the Ming Dynasty (1368–1644) and the Qing Dynasty (1644–1911).

During the mid-1500s, the Ming emperor Wanli, supported by the twenty thousand eunuchs and ten thousand female attendants who staffed his vast imperial compound in the Forbidden City, adopted silver as the imperial medium of exchange to pay administrative salaries, conduct interregional commerce, and collect taxes. Chinese merchants preferred Mexican pesos to other currencies, since each coin contained a precise amount of silver that had been carefully weighed by Spanish officials in Mexico. Sailing with the trade winds, the transpacific journey from Acapulco to Manila was the shorter of the two trips, while the far more dangerous eastward journey followed a northerly route, covering more than six thousand miles and frequently taking six to nine months.

The meals aboard the galleons were atrocious. More often than not, the rock-hard biscuits consumed on long journeys were infested with mealworms, so all passengers were entomophagists by default. Scurvy, a vitamin C deficiency stemming from a shortage of fresh fruits and vegetables, was a common shipboard ailment. Its causes were not understood at the time, so passengers and sailors heading to Mexico were unaware that their ship's cargo contained a nourishing remedy for this debilitating disease, pickled Chinese ginger, barrels of which had been packed in Manila for wealthy Spaniards across the ocean.

After the heavily laden galleons completed their return voyage to the Americas, the extravagant goods in their holds fetched high prices in Mexico City and Potosí. Insect products were among the most coveted of these items. Historian William Lytle Schurz described these cargoes in lavish detail: "Above all, save for a few years, [the Spanish galleons that sailed from Manila to Acapulco] were silk ships. Silks in every stage of manufacture and of every variety of weave formed the most valuable part of their cargoes. There were delicate gauzes and Cantonese crepes, the flowered silk of Canton, called *primavera* or 'springtime' by the Spaniards, velvets and taffetas and the *nobleza* or fine damask, rougher grosgrains, and heavier brocades worked in fantastic designs with gold and silver thread." As the success of the *china poblana* proved, an exotic material woven from the tensile threads of domesticated caterpillars half a world away could make a lasting impression on Latin American culture.*

The deep history of global interconnectedness is as much a product of environmental factors as it is a result of cultural forces. The historical record is replete with examples of how atmospheric changes in one location have dramatically altered weather patterns in far-off places. The most notorious of these "teleconnections," as meteorologists call such phenomena, is the El Niño–Southern Oscillation (ENSO). This periodic variation in the winds and sea-surface temperatures over the tropical Pacific Ocean off the coast of Peru can trigger extreme weather, from floods to droughts, elsewhere on the planet. The historian Mike Davis has even connected El Niño events and the contemptuous responses of European imperialists to the widespread famines that swept across the globe during the last third of the nineteenth century. Prior to these devastating episodes, of course, climate events in one part of the world were altering lives on opposite sides of the planet. Yet these linkages were often unpredictable.

Even the shrewdest predictor could not have forecast how a South American volcanic eruption would shape Europe's silk industry. When the Huaynaputina volcano erupted in southern Peru on February 19, 1600, it discharged so much sunlight-reflecting sulfate aerosols into the stratosphere that much of Eurasia experienced an unusually long winter. The delayed onset of Italy's springtime meant that mulberry trees did not unfurl their leaves in time for the silkworm-hatching season. Consequently, the larvae starved, local silk production plummeted, and Italian silk prices skyrocketed. As the agent for a major British textile importer wrote to his employer on June 15, 1600, "By all likelihood [silk prices] will not fall this year by reason of this cold summer, which so much hinders the springing forth of the leaf that they write out of Italy there are no mulberry leaves for the silk worms to feed on, so that silk will be very scant." With silk already a global industry by the onset of the 1600s, local silk production—and, hence, worldwide supply—could be affected by distant events.

---

* The Chinese counterpart to Mexico's *china poblana* is known as the *qipao* (旗袍) in Mandarin (and the *cheongsam* in Cantonese). This slender silk sheath, featuring short sleeves and a slit up one side from ankle to thigh, epitomized the Shanghai fashion scene in the 1920s and 1930s.

Indeed, climatic anomalies were only one of many dynamics that could upend commercial relationships in the fragile seventeenth-century global economy. Closer to home, religious persecution in continental Europe offered a stimulus to Britain's fledgling silk industry. In 1685, Louis XIV revoked the 1598 Edict of Nantes, thereby ending legal recognition of Protestantism in France. Subsequently, an influx of French religious refugees brought their silk-making expertise to Britain, enhancing silk production there and disrupting it in France. English silk manufacturers at Blackfriars in Canterbury and Spitalfields, in London, cordially welcomed the refugees, known as Huguenots.

Like their imperial rivals, the British also encouraged silk production in their overseas colonies. Despite their ambitions, these enterprises did not always flourish. In 1612, Captain John Smith described the tragedy that befell the first attempt to establish a silk-making operation in North America: "By the dwellings of the Savages are some great Mulberry trees; and in some parts of the Countrey, they are found growing naturally in prettie groves. There was an assay made to make silke, and surely the wormes prospered excellent well, till the master-workman fell sicke: during which time, they were eaten with rats." Britain's colonial sericulture scheme never fully recovered from this queasy episode. Despite patronage from the Stuart monarchs, the Virginia Colony lacked the skilled workforce to raise *Bombyx mori*. In addition, silkworms took poorly to North America's native red mulberry (*Morus rubra*) trees. Instead, the more profitable pursuit of tobacco (*Nicotiana tabacum*) cultivation quickly monopolized Tidewater planters' investments.

Farther north, New England colonists also experimented with sericulture. Often, their efforts were ambitious bids to achieve financial independence. The historian Edmund S. Morgan vividly described New England sericulture as "a kind of El Dorado to lure the prudent and industrious as surely as others were lured by a fat horse or a legend of gold in the hills." Eighteenth-century Congregationalist minister Ezra Stiles, the seventh president of Yale College and one of the founders of Brown University, was among those lured by the promise of silk. Stiles began raising silkworms in 1758 and learned to recognize many of his larvae individually. He dubbed his two largest caterpillars General Wolfe (after the famous British Army officer) and Oliver Cromwell (for the English military and political leader). A few decades later, Stiles founded a company to promote sericulture among New England's church parishes. At his behest, dozens of ministers planted mulberry trees and distributed saplings to members of their congregations. This silkworm craze peaked in the late 1830s and fell just as precipitously with a market crash and the onset of a mulberry blight in 1844 that devastated the region's trees.

New England's midcentury silkworm-and-mulberry obsession coincided with a growing awareness of the need to restore fertility to agrarian landscapes. While men and women on North America's colonial plantations and pastures may have found novelty in "larding the lean earth"—to borrow Shakespeare's fitting phrase—farmers in other parts of the world have long understood that excrement is a precious link in this loop. Manure and compost are troves of vital elements such as nitrogen, phosphorus, potassium, calcium, magnesium, and sulfur. These macronutrients are among the most important factors fostering plant growth. The Roman writer Lucius Columella (4–70 C.E.) counseled farmers to "bring as food for newly ploughed fallow ground whatever stuff the privy vomits from its filthy sewers." Similarly, a Chinese imperial treatise from 1737 observed, "The southerners accumulate nightsoil in pits. They treasure nightsoil as if it were gold." Animal droppings even nurtured close relationships between pastoralists and farmers. For example, in precolonial East Africa, manure from the cattle raised by Hinda clans fertilized the banana groves tended by neighboring Haya planters.

The notion that a civilization would ever willfully squander the wealth of its "privy vomits" produced great consternation among Europe's literary luminaries. In his 1862 novel, *Les Misérables*, Victor Hugo bemoaned the squandering of Parisian night soil: "A great city is the most mighty of dung-makers. Certain success would attend the experiment of employing the city to manure the plain. If our gold is manure, our manure, on the other hand, is gold. What is done with this golden manure? It is swept into the abyss." China, whose resilient civilization Hugo greatly admired, rarely squandered its fecal treasures.

Over many millennia, the Chinese had elevated manure recycling to an art. In medieval China, silkworm cultivation exemplified this finely calibrated approach to waste management. Developed during the Ming Dynasty by the peasants of Guangdong—known to most English speakers as Canton—the "mulberry embankment and fish pond" system (*cang ji yu tang*) used silkworm excrement and leaves from mulberry trees as food for pond-raised carp (*Cyprinus carpio*). In return, fish waste and decomposing organic matter provided rich fertilizer for the mulberry groves. Throughout southern China's Pearl River Delta, this aquaculture scheme yielded a sustainable nutrient cycling loop. Additionally, it offered rural families a source of dietary protein and a marketable commodity.

Although such strategies appeared to be organic outgrowths—both literally and figuratively—of the countryside, the Chinese state actively promoted them. Government officials frequently encouraged peasants with silk-related subsidies. In the early 1900s, the reform-minded Governor-General Xiliang established sericulture bureaus throughout Sichuan Province in southwestern China, providing farmers with mulberry trees, land, and sericulture training. Such extensive commitments of government resources demonstrated silk's standing as China's most valuable trade good.

Silk's commercial resilience was matched by its physical durability. Worlds away from rural China, the hardiness of silk manifested itself in one of the worst naval disasters in Britain's peacetime history. On August 29, 1782, the flagship one-hundred-gun HMS *Royal George* lay at anchor just beyond the entrance to Portsmouth Harbour, on the southern English coast. Hundreds of men, women, and children were aboard, visiting the crew, selling wares, and carrying out repairs. The ship was to sail in two days to join the British fleet in the Mediterranean when the foreman of a work squad felt it necessary to heel over the vessel to fix a normally submerged cistern pipe. This involved shifting many of the heavy cannons and recently loaded rum barrels toward the port side of the *Royal George,* which caused the starboard side to rise several feet out of the water. An unforeseen gale swept in and quickly submerged the open portals on the lower deck of the severely tilting vessel, inundating the ship with water. In a matter of minutes, the *Royal George* sank like a boulder. At least nine hundred people, including three hundred women and sixty children, drowned.

Despite a valiant rescue operation, only 255 survivors reached shore. Fifty-eight years later, the British Army's Corps of Royal Engineers

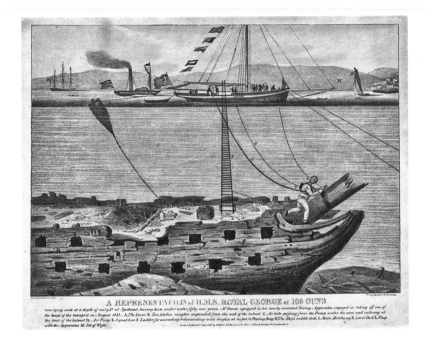

A REPRESENTATION of H.M.S. ROYAL GEORGE of 108 GUNS

A depiction of a nineteenth-century salvage operation to recover the
contents of HMS *Royal George,* which sank off the southern coast
of England on August 29, 1782, killing at least nine hundred people.
Among the items recovered nearly sixty years after this colossal
maritime disaster were articles of clothing fashioned from silk fabric
that had remained intact for more than half a century.

salvaged a cache of artifacts from the wreck. *The Times* of London
reported on October 12, 1840: "Next to brass, *the most durable article
found has been silk;* for besides pieces of cloaks and lace, a pair of black
satin breeches, and a large satin waistcoat with flaps, were got up, *of
which the silk was perfect,* but the lining entirely gone, as well as the
buttons, from the thread giving way. No articles of dress of woolen
cloth have yet been found, and therefore we may presume that they
have all decayed."

Silk also proved its lasting worth in the fashion world. For the British, the elegant fabric has long served as an emblem of social distinction. This cultural function was remarked upon by Lord Illingworth, a
witty and flirtatious man of means in Oscar Wilde's 1893 play *A Woman
of No Importance:* "A well-tied [silk] tie is the first serious step in life."
What Wilde may not have known was that this style-savvy transfigura-

tion required more than a hundred cocoons, painstakingly unraveled by the laborers who tended the cauldrons and worked the looms of their insect industry. The silkworm pupae whose threads eventually intertwined as the warp and weft of the dandy's cravat had devoured at least a thousand pounds of mulberry leaves. Wilde, a writer whose work is often hailed as a bridge to modernity, built his fashion sensibilities upon an insect institution with ancient roots.

As Lord Illingworth's advice demonstrated, silk has not always been an egalitarian textile. The sporadic passage of sumptuary laws—regulations designed to restrain luxury and prevent commoners from accessing status symbols—severely limited silk consumption. In 1567, Charles IX of France forbade all women but princesses and duchesses to wear silk. Such directives were not unprecedented. Roman emperors, Ottoman sultans, and various rulers of ancient China and Japan were among the many officials who restricted silk consumption for economic and symbolic reasons. Silk signified power, serving as a visual marker of class distinction. However, such divisions were hardly permanent. Reflecting upon this transient nature, a French manufacturing official wrote in 1886, "No industry is more dependent on fashion than that of silk fabrics." The world over, the fortunes of *Bombyx mori* ebbed and flowed with volatile consumer trends.

A few decades prior to his countryman's pronouncement about the volatility of style, a French refugee made an ill-conceived and calamitous attempt to bolster the fortunes of North American sericulture. In 1852, Étienne Léopold Trouvelot, a young artist with Republican leanings, escaped to the United States in the wake of Prince Louis-Napoléon Bonaparte's 1851 coup d'état. Trouvelot, his wife, and their two children fled the repressive political environment of midcentury France and eventually moved to the Boston suburb of Medford, Massachusetts. Trouvelot supported his young family by working as a scientific illustrator. A true Victorian autodidact, he joined the Boston Society of Natural History and wasted no time becoming a respected amateur entomologist. Trouvelot soon found himself intrigued by the possibilities of silkworm cultivation, and he began raising giant silk moths (specifically, *Antheraea polyphemus*) under a mesh canopy on his five-acre property. The very first issue of *The American Naturalist*, published in 1867, featured Trouvelot's testimonials about his wide-ranging entomological experiments: "For over six years I have been engaged in raising the Polyphemus worm, and here present the following imperfect sketch of the progress made from year to year in propagating and domesticating these insects from the wild stock." Trouvelot's "infant industry," as he called it, soon involved more than a million larvae.

Trouvelot's ulimate goal was to improve the hardiness of the larvae available to North America's sericulturalists. To advance this objective, he returned from a trip to Europe in the 1860s with a wooden crate full of live gypsy moths (*Lymantria dispar*). Trouvelot hoped that by breeding his precious cargo with American silk-producing moths he could produce a stronger, disease-resistant hybrid. One summer day in 1868 or 1869 (the record is unclear) several of the exotic creatures escaped from Trouvelot's Medford yard. These winged fugitives quickly made a feast of Medford's foliage. Within a decade, European gypsy moths were devastating tree canopies statewide. Maria Elizabeth Fernald, a prominent entomologist at the Massachusetts Agricultural Experiment Station, was the first scientist to identify the culprit and lead the eradication campaign.

The implications of Trouvelot's mishap were not lost on nineteenth-century commentators. In 1893, the writer Alice Bailey Ward mused, "A cluster of tiny eggs had been blown out of Mr. Trouvelot's window, and, like the dust on the palm of Moses, lifted and whirled, had brought forth a plague." The gypsy moth is indigenous to parts of Europe, Asia, and North Africa. Its larvae feed on the leaves of more than five hundred plant species, its undiscriminating palate making it one of North America's most destructive pests. During the gypsy moth outbreak of 1981, hordes of caterpillars dripped off trees, devouring at least nine million acres of forest canopy from Maine to Maryland. Vast stretches of deciduous forest across the eastern United States have been denuded by this invasive species.

Ironically, Trouvelot's scientific career did not flourish until the years that followed the gypsy moth debacle. Abandoning the insect world, Trouvelot turned to the heavens, eventually achieving widespread fame in astronomical circles. His seven thousand mesmerizing drawings of celestial objects and the fifty scientific papers he published on planetary research earned him an invitation to join the staff of the

Harvard College Observatory in 1872. Craters on both the moon and Mars bear Trouvelot's name. Mystified by moths, he found redemption in the stars. His tremendous accomplishments notwithstanding, Trouvelot's story succinctly proves a point made by the twentieth-century astronomer Martin Rees: "What makes things baffling is their degree of complexity, not their sheer size . . . a star is simpler than an insect."

We are accustomed to looking skyward for radical shifts of perspective: the angelic revelations of the Judeo-Christian tradition, the dawn of the sixteenth-century Copernican Revolution, the 1968 *Apollo 8* photo of Earth ascending over the moon's horizon. Yet insight also resides closer to the ground. Silkworms offer a convincing reminder of the abiding role of insects in everyday life. In our highly integrated contemporary world, it is now possible to ask, Who has not touched silk?

The capacity of silkworms to enhance human sensitivity to planetary concerns has long been acknowledged by certain cultures. In 1940, the Chinese writer Chiang Yee (Jiang Yi) fondly recalled his childhood in the Yangtze River city of Jiujiang, where he raised silkworms as pets: "Like other members of my family, particularly girls, I fed silkworms as a hobby. . . . From looking after silkworms we should acquire, it was felt, skillful hands and careful minds for dealing with bigger things."

In turn, silkworms have frequently been lavished with terms of endearment. The Japanese poet Kobayashi Issa (1763–1828) gives us a taste of this fondness in the following lines:

> They were called "Sir" [*sama*]
> when they were being raised—
> these silkworms were.

In Japanese, *sama* (様) suggests respect, and even affection. Yet there is also finality, and even a touch of nostalgia, in Issa's wording. The poem is set in the past tense, suggesting a closing act, the end of a pampered life.

The approach of World War II, which had so dramatically altered the shellac trade, also forecast closure for the long-standing human-silkworm relationship. Suddenly, silk became more difficult to obtain. In the United States, the army and navy used the limited supplies of the resilient fabric for manufacturing parachute canopies. The drive to find substitutes for Asian silk was intertwined with propaganda battles on the home front. In a 1938 pageant at Wardman Park Theater in Washington, D.C., the League of Women Shoppers advertised its ambition to "reveal the chic a woman can acquire without a thread of Japanese silk." Likewise, DuPont introduced nylon stockings to American consumers during a wave of anti-Japanese publicity. The editors of *DuPont Magazine* even suggested calling the May 15, 1940, nationwide debut of their new product "N-Day."

DuPont chemist Wallace Carothers first synthesized the polymer nylon on February 28, 1935. A 1940 *Fortune* magazine article hailed the fabric as the harbinger of miracles to come: "It is an entirely new arrangement of matter under the sun, and the first completely new synthetic fiber made by man." Yet in the long run, this alchemical marvel of modern science did not spell the end of the road for silk. In fact, it was merely a temporary stop along an enduring journey.

The properties of natural silkworm silk are surprisingly difficult to imitate. A 1998 study in the *Journal of Applied Polymer Science* noted, "Biological materials often show a combination of properties that cannot be reproduced by artificial means. Silks, produced either by spiders or moth larvae, are a good example when compared with artificial organic fibers." Many insects (caddisworms, blackflies, katydids, lacewings, sawflies, and fungus gnats) and other arthropods (mites and spiders) produce silk, but few achieve the high tensile strength and smooth appearance of domesticated silkworm threads.

Silk is one of many natural mysteries that humans have yet to unravel. In 1874, a British primer for youngsters noted, "Did you know, before, that a worm spins all the beautiful silk of the world; that no man has yet lived, wise enough to produce a substance in any other way, that will compare with it for beauty and durability?" A century and a half later, this assertion remains valid. As the anatomist James V. Lawry wrote in 2006, "Silk drives engineers to tears. Even though we can sequence silk genes and can splice these into the DNA of goats and bacteria, synthetic silk is still not mass-produced nor is it a high-strength material." Even today, Madame Moth continues to beguile her audience.

COCHINEAL

Piracy and entomology make an odd couple. But the notion of a bug-obsessed buccaneer is not as fanciful as it might seem. In fact, one such character played a starring role in the historical drama of the cochineal insect (*Dactylopius coccus*). Had Nicolas-Joseph Thiéry de Menonville managed to live a few years longer, he might have pulled off a biological heist with far-reaching consequences. Instead, the French biopirate succumbed to "malignant fever" in 1780 at the age of forty-one. Thiéry never witnessed the success of the French foray into the cultivation of cochineal, the tiny bug that supplied Europe's most coveted red dye for three centuries following the Spanish conquest of the Americas.

Before his untimely death, Thiéry produced a remarkable account of his journey into the heart of colonial Spanish America in pursuit of the valuable insect. His *Voyage à Guaxaca* ("Voyage to Oaxaca") reads like a spy thriller, brimming with the intrigues of international espionage in the Age of Empire. With sponsorship from the French Ministry of the Navy, Thiéry—the "new Argonaut," as he styled himself—hatched a scheme to abscond from Mexico with a supply of cochineal bugs and break the Spanish Empire's 250-year-old monopoly on the profitable transatlantic trade in carmine dye.

At the time, Spaniards vigilantly guarded their commercial dominance of red pigments. The French paid dearly for the cochineal they purchased from their European neighbors. Each one-pound block of dye, handmade by peasants in rural Mexico and shipped across the Atlantic, contained the crushed bodies of seventy thousand female cochineal insects. When dissolved in an alkaline solution, these bricks yielded stunning crimson, scarlet, and purple hues for coloring threads and tinting paints.* Much of this pricey dye went directly to the renowned Gobelins tapestry works, established by Louis XIV in Paris in 1662 to supply opulent furnishings for French royal palaces. Frustrated by their dependence on a rival empire, the French sought access to a cheaper source of red, a potent symbol of royalty and religious authority in Europe. The only route to this objective extended through the pueblos and plantations of southern Mexico.

Thiéry was compelled by the prospect of attaining glory for his country and the admiration of his peers. By all accounts, he was patriotic, brash, and single-minded, an embodiment of the traits needed for the grueling mission he would eventually undertake. Born in Lorraine in 1739 to a family of prominent judges and lawyers, Thiéry dutifully entered the family business. His real passion, though, was natural history. Soon after earning his law degree, he escaped to Paris and began studying under the most eminent French botanists of the time. At the Jardin du Roi (the King's Garden), he mastered the intricacies of plant taxonomy. He also discovered the works of the Enlightenment writer Guillaume Thomas François Raynal, who bemoaned the exorbitant cost of cochineal: "Its price, which is always very high, should have excited the emulation of the nations which cultivate the islands of America . . . however, New Spain has remained in possession of this rich production." Declarations such as Raynal's convinced Thiéry of his calling. He would cross the Atlantic, sneak into Mexico, steal cochineal for France, and single-handedly disrupt Spain's commercial stranglehold on the global trade in carmine dye.

Initially, Thiéry's family members mocked his outlandish plan, but Thiéry rejected their appeals to reason. Nothing would deter the stubborn young botanist. The French government endorsed Thiéry's

---

* There are two types of cochineal dye. The first derives from crushing the dried bodies of female *Dactylopius coccus* insects, yielding a final product of about 17–24 percent carmine dye per unit volume. The second is the product of controlled extractions from insect bodies using acidic, aqueous, and alcoholic solutions to produce a purer, darker carminic acid.

scheme and granted him a considerable stipend of six thousand livres (the equivalent of six thousand pounds of silver) to fund his heist. In 1776, Thiéry embarked for the Caribbean. After a sixty-six-day transatlantic crossing, which left him "tired, disgusted with the sea," he arrived at Port-au-Prince in the French colony of Saint-Domingue (later known as Haiti), which occupied the western third of the island of Hispaniola. It was the most profitable settlement in the West Indies. Valuable plantation crops of sugar, coffee, cocoa, indigo, and cotton thrived in Saint-Domingue's tropical soils, enriching a few thousand white planters, who cultivated decadent lifestyles from the backbreaking labor of nearly half a million African slaves. In 1791, little more than a decade after Thiéry's visit, the slaves of Saint-Domingue rose up against their masters. Within a few years, the insurgents had forced the French colonial administrators to emancipate them, paving the way for the struggles that gave birth to the nation of Haiti and the dissolution of slavery in the Americas.

After disembarking in Port-au-Prince, Thiéry convalesced while formulating a plan to sneak into Mexico and find his desired insects. He decided to pose as an eccentric Catalan doctor in search of the herbal ingredients to make an ointment for treating gout. Thiéry would use this pretense to hoodwink Spanish officials, venture deep into Mexico's interior, and fill his collecting cases with precious cochineal bugs and the bulbous pads of prickly pear cacti, on which these insects dwell and dine. He would then slip out of Mexico undetected and establish a nursery for the valuable creatures and their host plants in Saint-Domingue.

In the winter of 1777, Thiéry caught wind of a maritime salvage operation preparing to depart for Cuba. Not wanting to squander an opportunity to get one step closer to his goal, he secured passage aboard the brigantine *Dauphin,* which sailed for Havana on January 21. The young botanist hoped to elude Spanish officials, but he came prepared for any eventuality. In his waistcoat pocket, he carried the falsified papers of a physician. Thiéry's luggage, however, betrayed his true intentions. Among his few possessions were "a quantity of vials, flasks, cases, and boxes of all sizes."

Arriving in Havana, Thiéry immediately began mingling with the city's elites. The Frenchman flaunted his diamond ring and exhibited

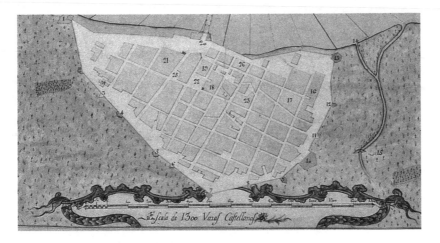

A map of the port city of Veracruz, Mexico, in 1777, the same year the
Frenchman Thiéry contrived an elaborate plan to capture cochineal bugs
there and smuggle them back to French colonial territory in an effort to
break the centuries-old Spanish monopoly on the deep-red dye
manufactured from the dried bodies of the insects.

elevated manners, attended the opera, dined with Spanish aristocrats,
and befriended the governor, all the while improving his meager com-
mand of the Castilian language. Thiéry's newfound associates were
delighted by the doctor's genial disposition and helped him secure a
passport to Veracruz.

By the time of Thiéry's arrival, Mexico's foremost east coast harbor
had a well-deserved reputation as Spain's gateway to the riches of the
Americas. In the words of Uruguayan writer Eduardo Galeano, there
was "a river of silver flowing to Europe through the port of Veracruz."
For the continent's indigenous inhabitants, it was a threshold of horrors, bringing disease, violence, and death. The Spanish conquistador
Hernán Cortés had arrived in Mexico in 1519. His forces and their
Native American allies had conquered the Aztec Empire two years
later, inaugurating a period of devastating epidemics and brutal colonial occupation for the region's indigenous inhabitants.

Thiéry disembarked at the Veracruz docks in late March 1777. He
received a warm welcome from the locals and was overjoyed to find
freshly churned pineapple ice cream for sale in the city's cafés. Despite
these auspicious beginnings, Thiéry and the governor of Veracruz,

Don Fernán Palacio, did not get off to a promising start. The Frenchman was taken aback by Don Fernán's "brusque tone" and "foul language." The governor reluctantly gave Thiéry temporary permission
to stay in the city and botanize, but he confiscated the Frenchman's
passport and ordered his guards to monitor the suspicious foreigner's
activities.

Thiéry carefully adhered to the alibi that he was a plant-obsessed
doctor from Catalonia. In short order, he endeared himself to many
citizens of Veracruz by identifying a local variety of jalap (*Ipomoea
purga*), a flowering plant in the morning glory family known for
stimulating the bowels. Prior to this rousing discovery, Veracruzanos
who suffered from constipation had imported jalap roots from the
town of Xalapa, more than fifty-five miles to the northwest. As Thiéry
proudly recorded in his journal, "Such a discovery gave me a reputation throughout the city . . . as an extraordinary man who knew how
to find treasures unknown to those who possessed them." Clearing
such an obstruction was not without its rewards.

Despite his impressive floral discovery, Thiéry remained mystified
about where to find cochineal. His fortunes shifted when he overheard
some merchants discussing various grades of the insect dye. These
men concurred that the finest-quality red colorings were obtained
from Oaxaca City, nearly three hundred miles to the south of Veracruz. This distant locale and its bugs would quickly become Thiéry's
toison d'or, his "golden fleece."

Unlike many residents of Veracruz, the Spanish colonial bureau-
cracy was not thrilled about Thiéry's extended stay in Mexico. After
the Frenchman applied for a permit to travel inland, the viceroy of
New Spain (the Crown's highest-ranking official in North America),
Antonio María de Bucareli y Ursúa, became suspicious. He declared
that Thiéry could no longer remain in Spanish territory and should
leave Veracruz in three weeks, aboard the next departing ship. When
an official read the viceroy's proclamation aloud to Thiéry, the Frenchman feigned bewilderment. Internally, he was tormented by the prospect of failure. How could he face his family, let alone his compatriots,
if he returned home empty-handed?

He resolved to attempt a perilous three-week mission to find the
objects of his entomological quest and their prickly pear hosts. Thiéry

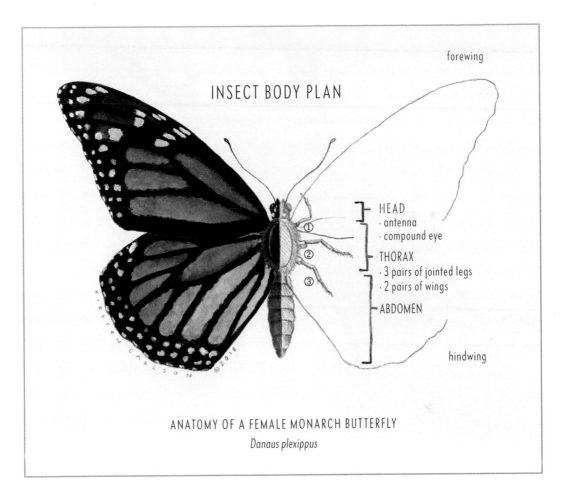

INSECT BODY PLAN

forewing

① HEAD
· antenna
· compound eye

② THORAX
· 3 pairs of jointed legs
· 2 pairs of wings

③ ABDOMEN

hindwing

ANATOMY OF A FEMALE MONARCH BUTTERFLY
Danaus plexippus

Insect body plan *(Prepared by Kirsten Carlson for Edward Melillo)*

Albrecht Dürer, *Stag Beetle*, 1505 *(Courtesy of the J. Paul Getty Museum)*

The Deadly Mantis movie poster, 1957

A tree branch with shellac secreted by *Kerria lacca* insects *(Courtesy of Jeffrey W. Lotz, Florida Department of Agriculture and Consumer Services, Bugwood.org)*

POBLANAS.

Nineteenth-century Mexican *poblanas* wearing dresses made with Chinese silk, as shown in Carl Nebel, *Voyage pittoresque et archéologique dans la partie la plus intéressante du Mexique* *(Paris: Chez M. Moench, 1836)*

Illustration of a man collecting cochineal from a nopal cactus, from José Antonio de Alzate y Ramírez,
Memoria sobre la naturaleza, cultivo, y beneficio de la grana, 1777, colored pigment on vellum (Courtesy of
the Newberry Library)

A bee pollinating a flower *(Courtesy of Friends of the Earth, United Kingdom)*

An "insect hotel" on a farm in northern Spain. Such habitats help conserve endangered insect species. *(Photograph by CADV17.)*

Ancient Egyptian beekeepers extracting honey from a hive (ca. 1479–1425 BCE). Painted at the Tomb of Rekhmire in Qurna, Egypt, by Nina de Garis Davies for the Egyptian Expedition of the Metropolitan Museum of Art, 1926 *(Courtesy of Metropolitan Museum of Art)*

Monarch butterflies (*Danaus plexippus*) overwintering in Santa Cruz, California, during 2007 *(Courtesy of Brocken Inaglory)*

vowed to return to Veracruz with the illicit cargo before his prear-
ranged exile. This dodgy adventure would require traveling nearly six
hundred miles in fewer than twenty-one days with neither a passport
nor a map. He would be a vagrant in a strange land where he spoke
only a muddled version of the colonizer's language and knew few, if
any, words of the many regional dialects he would encounter along the
way. Capture by Spanish soldiers would most certainly result in death,
but the prospect of triumph spurred Thiéry onward, at least according
to his melodramatic journal entries.

In the middle of the night, the day after the viceroy's proclamation,
Thiéry scaled the Veracruz city wall and plunged into the countryside. He fretted about his appearance: "Without a coat, I looked like
a stranger; a handkerchief and a large hat over my head scarcely reas-
sures me against the eyes of a multitude of curious people." Thiéry's
anxiety about being discovered caused him to employ every precau-
tion. He skirted toll gates, pushed onward through torrential rains,
painstakingly evaded Spanish patrols, and suffered for days without
food or shelter in order to keep moving. When he did find meals and
lodging, they were often in the hut of a "poor Indian," and he paid
dearly for the few eggs and tortillas his hosts procured. After several
days of grueling travel, he acquired a horse, hired a Native American
guide, and obtained detailed directions from a jovial Carmelite monk
who trusted Thiéry's assertion that he was a Catholic pilgrim seeking
the church of Nuestra Señora de la Soledad in Oaxaca City.

In Thiéry's elaborate rendition of events, the only temptation that
challenged his resolve during the journey was a beautiful woman who
ran a roadside café. She was "nearly naked, having only a flounced
muslin petticoat trimmed with a rose-colored cord, and a shirt which
left her shoulders bare." Thiéry confessed, "Her charms had a disori-
enting effect on my senses." He was sorely tempted to pay her a gold
piece for her favors, but his devotion to the task at hand intervened to
curb his cravings. He turned around and left without saying a word or
looking back.

A few days later, Thiéry reached the pueblo of Gallatitlán and finally
witnessed the object of his true desires. At a spacious hacienda, row
upon row of prickly pear cacti were encrusted with clusters of tiny
dots. Feigning innocence as to what he had seen, he visited the home

of the Native American owner and asked what purpose the plant served. "He responded, it is to cultivate the *grana*." As Thiéry was well aware, *grana* was the Spanish term for the female cochineal bug. This was the moment he had yearned for, but he remained concerned that "the tiny insects were covered in a white powder." Were these minuscule creatures really cochineal? Thiéry was relieved when he crushed one on a sheet of white paper and discovered that it bled "the true purple of kings."

Thiéry paid the farmer a few coins, took some samples of cacti and cochineal, wrapped them in towels, and departed for Veracruz. On his way, he and his guide picked up more prickly pears and bugs from rural cultivators, which they carefully packed into woven baskets for the journey north. They also found ripe vanilla pods, another prized Spanish commodity, which they mixed with a jumble of other botanical cuttings in Thiéry's collecting boxes to avoid detection.

The Frenchman and his treasured cochineal bugs made it back to Veracruz before his ship's scheduled departure for Saint-Domingue. No stranger to self-aggrandizement, Thiéry triumphantly recounted his evasion of "two viceroys, six governors, thirty *alcaldes* [mayors]" and twelve hundred Spanish customs officials and guards. Having run this gauntlet, Thiéry now boarded a ship. However, he would soon face a new danger.

During the stormy voyage across the Gulf of Mexico, several sailors caught Thiéry tending to his fragile cacti and his contraband insects. As the Frenchman dutifully explained to the Spanish mariners, the cochineal, cacti, vanilla, and jalap in his containers were vital ingredients for his proprietary gout remedy. To bolster his claims, Thiéry divulged to the interrogators that the top secret ingredients of his healing ointment also included incense, gold dust, and silver leaf. For extra flair, "I added that I mixed in some blessed linen that had touched the relics of Santo Torribio."

This elaborate ruse worked. The French smuggler had passed the final test in his lengthy ordeal. After Thiéry arrived at Saint-Domingue in September 1777, his countrymates celebrated his accomplishment. Louis XVI of France honored Thiéry with the title of *botaniste du roi* (the king's botanist). This appointment came with a generous annuity of six thousand livres. Thiéry established a nursery in Port-au-Prince for his transplanted cacti and cochineal bugs and sent samples of the plants and insects to the scientific academy at Cap Français, some eighty-five miles to the north. He had high hopes for a profitable cochineal industry in the French Caribbean that would serve as the engine of growth for a maturing, entrepreneurial class of freedmen known as the *gens de couleur* (people of color). However, within three years of his return to the warm embrace of the French Empire, Thiéry died in Port-au-Prince of "malignant fever."

René-Nicolas Joubert de la Motte, the French royal physician who assumed Thiéry's duties at the Port-au-Prince garden, lacked the zeal and attentiveness to care for the cochineal bugs. Hungry rats, ravenous birds, and predatory ants feasted upon Thiéry's cherished insects, and the Mexican prickly pear cacti that he had so laboriously smuggled out of Veracruz rotted in Hispaniola's heavy rains.

Not all was lost, however. A private colonist named A. J. Brulley discovered a wild variety of cochineal among the ruins of Thiéry's garden and used these hardy insects to establish a four-thousand-plant nursery on the northern side of the island. Brulley sent samples of his red dye to the Paris Academy of Sciences, where chemists determined that his product very nearly approximated the finest hues of Mexican cochineal. By the time the French had begun to develop self-sufficient cochineal production, the revolution of 1789 had transformed the national relationship to color. Scarlet hues and crimson tones of ecclesiastical vestments and kingly robes were hastily displaced by the more democratic red, white, and blue stripes of the *tricolore*, adopted as the French flag in the 1790s.

France's short-lived encounter with the cochineal insect was the outcome of longer-term developments in Latin American history. More than two millennia before Columbus crossed the Atlantic, peoples in the Andes were revolutionizing the human relationship to the color spectrum. The discovery of carmine-infused fabrics from Peru's Paracas culture (800–100 B.C.E.) demonstrates the protracted history of cochineal dyeing. As with silk, our contemporary knowledge of an ancient human-insect relationship comes from mummified corpses. In the 1920s, excavations near southern Peru's Ica Valley unearthed hundreds of entombed bodies wrapped in cloth mantles that had been dyed in radiant hues with plant-based pigments and vibrant red insect-derived cochineal colorings.

Following their early adoption in the Andes, the organisms and techniques of cochineal dyeing spread north into Central America. It is likely that mariners plying ancient oceanic trade routes along the Pacific coast brought the cochineal insect and its host cacti northward. The Mesoamerican Toltec culture (800–1200 C.E.) rapidly mastered the production of deep red dyes from this newly introduced insect.

The Aztecs, who inherited cochineal cultivation techniques from their Toltec predecessors, knew cochineal as *nocheztli*, or "blood of the prickly pear" in the Nahuatl language. In Mexico, two domesticated nopal cactus species, the "nopal de San Gabriel" (*Opuntia tomentosa* var. *hernandezii*) and the "nopal de Castilla" (*Opuntia ficus-indica*), have sustained and sheltered cochineal insects for many centuries. These cacti feature prickly barbs the size of sewing needles; edible, paddle-shaped pads (or cladodes) that Mexican cooks have traditionally chopped into soups, salads, egg dishes, and salsas; bulbous, sweet fruits that can be harvested and eaten after removing the small, hairy prickles; and eye-catching red, orange, yellow, and white flowers that bloom after the spring and summer rains. Prickly pear cacti have been pivotal in Mexican history. Legend has it that the Aztecs chose the site of their capital city, Tenochtitlán, on a swampy island in Lake Texcoco, after seeing a golden eagle perched atop an *Opuntia* cactus (where these birds often nest), clutching a rattlesnake in its talons. This scene is the centerpiece of the Mexican flag.

The Aztecs continued to hold prickly pear cacti and the cochineal bugs they hosted in high esteem. The Codex Mendoza, which contains elaborate records of the use of honey as tribute in the early 1500s, also depicts Aztec emperor Montezuma II accepting sacks of dried cochineal and carmine-dyed cloth as treasured offerings from his subjects. The Spanish conquistadors who arrived in Mexico in 1517 were awestruck by the advanced nature of Aztec dyes. In 1520, Hernán Cortés wrote to his patron Holy Roman Emperor Charles V (who was also King Charles I of Spain), "They have colors for painting of as good quality as any in Spain, and of as pure shades as may be found anywhere." The most important source of these "pure shades" was the cochineal bug.

Despite widespread misconceptions, the cochineal bug is not a beetle. It is a scale insect of the order Hemiptera, making it "a true bug," according to entomologists. The cochineal insect has piercing-sucking mouthparts and is a close relative of aphids, cicadas, and the lac bug

Indigenous farmers in colonial Mexico harvesting cochineal insects from the paddles of a large nopal (prickly pear) cactus in the mid-1600s. Despite Spanish attempts to reorganize and expand red dye production from the sixteenth century onward, the rural practice of cochineal cultivation remained in the hands of local farmers, who continued to use traditional techniques that predated European conquest.

Kerria lacca. The sought-after red dyestuff of the *Dactylopius* insect comes from the carminic acid that the wingless female secretes to poison her predators. The females attach themselves to the pads of the cacti and use their hollow, strawlike proboscies to suck moisture and nutrients from the plants. In the process, they ooze strands of white, cottony wax to protect their bodies and egg masses, much as lac tunnels and silkworm cocoons shelter those insects' young at vulnerable life stages. Writing in 1653, the Spanish Jesuit Bernabé Cobo described the full-grown female cochineal insect as *"la grandeza de un garbanzo ó frísol"* ("the size of a chickpea or kidney bean").*

Like lac cultivation and sericulture, the task of nurturing cochineal

* The spellings of some seventeenth-century Spanish words differ from those of today.

bugs demanded tireless devotion to both the insects and their host plants. The female cochineal bug is virtually defenseless because of her immobility, so cochineal farmers—called *nopaleros,* following the Spanish name of the host cactus—spent many hours each day warding off predators, which included wild cochineal insects (known in Spanish as *cochinilla silvestre*), the telero worm (*Laetillia coccidivora*), chickens, turkeys, lizards, woodpeckers, mice, and rats.

In preparation for cochineal production, nopal cacti require between two and three years to mature. Cultivators then constructed tubular nests from corn husks or interwoven palm leaves, each housing several dozen egg-laden female bugs. Nopaleros used wires to affix these finger-sized containers to the nopal pads. Once the insects hatched and developed for four or five months, farmers harvested the bugs and drowned or steamed them, before drying them in the sun. These processes reduced an insect's volume by two-thirds.

Traditionally, after collection and desiccation, nopaleros ground the dried bugs (*grana seca*) with a stone rolling pin (*metate*) and packed the dye into leather sacks (*zurrones*) for transportation. Spanish colonizers did little to disturb elaborate local cochineal cultivation techniques, preferring instead to extract enormous profit from cochineal exports.

Imperial officials in New Spain saw the virtues of this lucrative trade. At the same time, they worried about its local economic repercussions. Indigenous farmers frequently interspersed *Opuntia* cacti and cochineal bugs among their corn, bean, and squash plants, much as silk producers and lac cultivators integrated the cultivation of insects with garden food crops. On occasion, agriculture and cochineal husbandry competed for scarce resources. In 1553, members of the *cabildo* (the municipal council) of Tlaxcala, in Mexico's east-central region, fretted that too much land and labor were devoted to dye production: "Everyone does nothing but take care of cochineal cactus; no longer is care taken that maize and other edibles are planted." While these concerns seemed urgent at the time, they were soon overshadowed by the undeniable profitability of the cochineal business.

In the early 1600s, the Spanish Carmelite monk Antonio Vázquez de Espinosa saw nothing but financial opportunities in Tlaxcala's cochineal: "[Tlaxcala] takes in quantities of fine cochineal, as do other cities and villages in its jurisdiction; and if the Indians paid tithes on it, as the bishop proposes and has taken legal steps to authorize, the diocese will have an annual income equal to that of the archdiocese of Toledo." Taxation proved complex, however. Few Spaniards understood the labor-intensive process of cochineal cultivation, and economies of scale rarely enhanced output. Thus, small operations, often funded with credit from large landowners, prevailed. According to the viceroy of New Spain, some twenty-five to thirty thousand residents of the southern state of Oaxaca (approximately one-third of the households) produced cochineal during the mid-1790s.

From the very beginning of Spain's transatlantic conquest, cochineal had enjoyed spectacular success in European markets. Following the defeat of the Aztec Empire (1519–21), Cortés sent samples of cochineal to his patron, Charles V. Within a few decades, imports of rich scarlet dye from the Americas had made vibrant impressions across Eurasia. In the sixteenth-century Florentine Codex, the Spanish-born Franciscan missionary Bernardino de Sahagún wrote, "This *grana* is known in this land and beyond her shores, and there are great testaments to it; it has reached China and the Ottoman Empire."

The arrival of Latin American cochineal in the Far East was among the many lasting outcomes of a phenomenon that the American environmental historian Alfred W. Crosby famously called the Columbian Exchange. This extensive transfer of flora, fauna, and microorganisms between the Americas, on one side, and Eurasia and Africa on the other, followed Christopher Columbus's 1492 arrival in Hispaniola, the island that is now home to Haiti and the Dominican Republic. In a watershed moment of transatlantic interchange, crops native to the Americas—tobacco, potatoes, tomatoes, maize (corn), and cacao (the bean from which chocolate is made) moved east and invigorated Old World diets. A few diseases, like syphilis and polio, also crossed from the Americas to Europe. However, it was the westward passage of smallpox, measles, malaria, yellow fever, influenza, and chicken pox from Europe that had the most severe consequences for the civilizations of the New World. Upwards of fifty-six million people throughout the Americas perished from the epidemics that accompanied European conquest.

Crosby contended that the fifteenth-century encounter between the

Old and the New Worlds "reknit the seams of Pangaea," the supercontinent that began to fracture some 175 million years ago. Voltaire's 1759 novella *Candide* foreshadowed this twentieth-century rendering of the biological invasion of the Americas. As the ever-optimistic Professor Pangloss quipped to his student, "If Columbus had not in an island of America caught this disease [syphilis], which contaminates the source of life . . . we should have neither chocolate nor cochineal."

Long before the Genoese navigator and his crew acquired syphilis, chocolate, and cochineal in the Americas, red had become the color of choice for European royalty. Its lasting associations with virility, sacrifice, and prestige made it among the most desirable hues on artists' palettes and weavers' looms. Prior to the arrival of cochineal, European dye makers manufactured their ruddy pigments from cinnabar (mercury sulfide), lichens, and madder plants (*Rubia tinctorum*). The Mediterranean sea snail (*Hexaplex trunculus*) and a pair of insects—Polish cochineal (*Porphyrophora polonica*) and kermes (*Kermes vermilio*)—also yielded scarlet dyes. However, once Mexican cochineal arrived on the scene, it proved superior to all of these because of its sheer vibrancy and fastness, its ability to retain color over time. Within half a century after reaching Spain in the 1520s, Mexican cochineal had replaced Polish cochineal and kermes as Europe's most widely used red dye. In 1599, the respected Mexico City resident Gonzalo Gómez de Cervantes noted that the Spanish were as eager for shipments of *grana fina*, or fine cochineal, as they were for cargoes of bullion. Indeed, by the eighteenth century, the carmine insect secretion was Mexico's second most valuable export, after silver.

Cochineal invigorated Europe's art world. Baroque painters expanded their palettes with the intense, saturated reds of cochineal-dyed paints. Michelangelo Merisi da Caravaggio's *The Musicians* (1595), Peter Paul Rubens's *Portrait of Isabella Brant* (1610), and Cristóbal de Villalpando's *Saint Rose Tempted by the Devil* (1695) are stunning examples of this red. Later, throughout the final years of the nineteenth century, painters like Paul Gauguin, Auguste Renoir, and Vincent van Gogh continued to enliven their canvases with cochineal hues. Using pigments from the crushed bodies of Mexican insects, Van Gogh achieved the luminescent burst of red at the center of his celebrated 1888 painting *The Bedroom*.

Cochineal even performed on the Shakespearean stage. When playing *Macbeth,* the esteemed nineteenth-century British actor William Macready relied on the vermilion insect secretion to imitate human blood. On at least one occasion, a lack of the necessary ingredient led to a rather unfortunate outcome for a bystander. Stage actor Clive Francis described how the scene unfolded at a Manchester theater:

> Macready stumbled off stage one night, only to find that the bowl of blood (cochineal), for him to smear on his hands, had failed to be set. In a blind panic, Macready rushed up to an inoffensive commercial traveler, who was standing in the wings, and without warning, punched him violently on the nose. Blood spurted everywhere. "Forgive me," hissed Macready, rinsing his hands under the man's nose, "but my need is somewhat urgent."

Like Macready's unwary victim, most Europeans were innocent spectators to the cochineal-tinted drama playing out across the world stage.

As with shellac and silk, Europe's consumers were mystified for centuries about the famed dye's origins. Throughout the sixteenth and seventeenth centuries, European writers disputed whether this coveted carmine coloring came from insects, worms, berries, plant seeds, or vegetable leaves. In part, the confusion stemmed from the common Spanish name for the dye's source, *grana,* which literally translates as "grain." In 1555, the English observer Robert Tomson declared, "The Cochinilla is not a worme, or a flye, as some say it is, but a berrie that groweth upon certaine bushes in the wilde fielde, which is gathered in the time of the yeere, when it is ripe." Yet 133 years later, little had changed. In 1688, the Royal Society of London published an account titled "Concerning Cochineel, Accompanied with Some Suggestions for Finding Out and Preparing Such Like Substances Out of Other Vegetables." The secrecy surrounding cochineal persisted. For more than two and a half centuries, Spain banned the export of live insects, ensuring its monopoly on the cochineal trade and generating an air of mystery around the dye's origins and manufacture.

Thiéry de Menonville's brazen act of biopiracy proved that cochineal cultivation had a future outside Spanish America, yet the Frenchman's heroic efforts had not weakened Spanish dominance over the profitable insect commodity. British cultivators enjoyed even less success than their French counterparts. Vying for financial rewards pledged by the British East India Company, Britons established cochineal farms in South Africa, India, and Australia. All three projects failed.

The results were particularly devastating in Australia, where *Opuntia* cacti introduced in 1788 proliferated beyond the bounds of plantations; several species became damaging and invasive on a continent where they had few natural predators. Farmers could not clear land choked with towering "cacti forests," and many attempts at eradication proved futile. In an ironic turn of events, insects came to the rescue. During the 1920s, scientists working for the Australian government introduced billions of South American cactoblastis moths (*Cactoblastis cactorum*) to vast, untamed expanses of prickly pear cacti. Within a decade, cactoblastis moth larvae, which contentedly munch on nopal cactus pads, had devoured and destroyed millions of acres of the invasive plant. In contrast to Étienne Léopold Trouvelot's calamitous attempt to strengthen North American silk moths by crossbreeding them with European gypsy moths, the Australian entomologists who transported cactoblastis moths across the Pacific from South America achieved their objectives.

Because of repeated failures to replicate cochineal production outside Spanish America, the dye's value tended to be high in Europe. Among British military ranks, a hierarchy of lac and cochineal insect secretions prevailed. A popular story from the mid-1800s highlighted this distinction: "The red coats of the British soldiers, meaning common soldiers, are all coloured with the inferior sorts of lac-dye. As for the officers, whose cloth is a good deal more brilliant, they are painted up with cochineal from Mexico." Once again, several insect commodities met in the realm of imperial spectacle.

As Europe's textile manufacturers learned, cochineal formed a more complete bond with cloth in the presence of a mordant, or dye fixative. Artisans often used the mineral salt aluminum sulfate to achieve this result. Such practices built upon traditional Native American cloth-coloring techniques from the highlands of what are now Peru, Bolivia, and Ecuador. Long before the Spanish invasion, Andean craftspeople had used acidic and alkaline minerals to achieve a range of cochineal-dyeing effects and an assortment of resplendent hues, from dazzling roses to deep purples and from sunset oranges to creamy pinks.

In the seventeenth century, Spanish Jesuits—cosmopolitan missionaries who divided their attentions between spreading Roman Catholicism and chronicling the rich Native American cultural traditions they encountered—conveyed these coloring practices to Europe's textile makers. In the process of acquiring techniques and botanical know-how from artisans in the Americas, Jesuits learned of three other plant-based dyes and eventually brought these with them across the Atlantic: indigo (*Indigofera tinctoria*), a species from the bean family that yields a deep blue pigment; campeche wood, or *palo de campeche* (*Haematoxylum campechianum*), which, when combined with various metals, imparts colors ranging from dark blue to black hues; and brazilwood (*Paubrasilia echinata*), from which producers extract a red dye called "brazilin." The transatlantic transfer of this traditional ecological knowledge from indigenous American communities to colonial metropoles was an underappreciated forerunner to the artistic revolutions that swept across post-Renaissance Europe.

Wanting to expand their ability to supply their European neighbors with fashionable red dyes, Spaniards eventually established cochineal operations beyond the Americas. The nineteenth-century English traveler Charles Edwardes reported about one such venture on the Canary Islands, a Spanish archipelago off the Atlantic coast of northwestern Africa: "The insect was not introduced into Tenerife [the largest of the Canaries] until 1825; and for a time it could not be encouraged to propagate successfully. A priest was the discoverer of the right method of nurture, and to him it is due that from 1845 to 1866 an annual crop of from two to six million pounds of cochineal was produced." This transatlantic transfer of natural cochineal production coincided with a revolution in synthetic chemistry. In 1858, William Henry Perkin, an assistant to German chemist August Wilhelm von Hofmann, invented a man-made purple pigment known as mauve. The compound, extracted from coal tar, was the world's first aniline dye.

Mexico's ambassador to the United States, Matías Romero, captured the hubris of the Synthetic Age when he wrote, in 1898, "But recent discoveries in chemistry have supplied other substances for dyeing which are very cheap, especially aniline, and the price of cochineal has fallen considerably, so that now it is hardly raised at all." Obituaries for cochineal abounded during the following decades. As a Virginia newspaper headlined one of its 1912 stories, "COCHINEAL IS NEAR END: SOON TO BECOME THING OF HISTORY LIKE TYRIAN PURPLE OF ANTIQUITY." Yet, much like the brief vanishing acts of silk and shellac, the disappearance of cochineal proved short-lived.

Carmine extract (to use one of the many names for insect-derived cochineal) has become ubiquitous once again. This time, however, it is not in the vestments of popes and kings but, rather, in common foods, where it now reigns supreme as a colorant. Mock crab legs, flavored waters, berry yogurts, Ruby Red grapefruit juice, high-end coffee drinks, and craft cocktails feature cochineal among their ingredients. It appears on labels as carmine, carminic acid, Natural Red 4, C.I. 75470, or E120.

Government regulations, stimulated by scientific evidence and widespread consumer protests, drove this shift. In 1990, the FDA responded to tests linking Red Dye No. 3, one of the oldest and most widely used synthetic food colorings, to thyroid cancer in rats and banned it. The nontoxicity, the chemical stability, and the relatively low price of cochineal have made it a desirable replacement for such synthetics. As the website of a Kentucky-based supplier of the insect secretion noted in 2013, "Carminic acid is one of the most light and heat stable of all the natural colorants and is more stable than many synthetic food colors. . . . Derivatives from cochineal are increasing in use due to the influence of the 'natural' trend." At times, certain culinary applications have drawn public scrutiny. In April 2012, Starbucks switched from cochineal to a vegetable-based dye in its strawberry "Frappuccino" after consumers protested the presence of bugs in their beverages. While Mexican farmers still supply carmine to the global market, 85 percent of the world's cochineal now comes from the highlands of Peru, where more than one hundred thousand families are involved in its production.

The past, present, and future of cochineal were on vivid display during the summer of 2015 when New Mexico's Museum of International Folk Art organized an exhibition called "The Red That Colored the World." The show featured an international collection of 130 cochineal-tinted objects. These items, on loan from private holdings and museums from as far away as Iran and China, chronicled the global spread of cochineal in the sixteenth century, its decline in the late 1800s, and its revival in the twentieth century. Ranging from embroidered altar cloths and stylish silk evening gowns to luxuriant Renaissance paintings and a Lakota courtship blanket woven from strands of Spanish thread, the pieces illustrated the remarkable diversity of ways in which humans around the world have used cochineal bug secretions to create objects of beauty.

Contemporary artists continue to draw inspiration from cochineal. The Mexican-born painter Elena Osterwalder combines it with other pre-Hispanic materials to create stunning compositions. Her 2015 installation in Columbus, Ohio, *Red Room,* blended cochineal-infused paints with traditional Mexican *amate* paper made from pounded tree bark. Osterwalder's exhibit displayed dozens of large, square panels, lit from various angles to showcase a vast spectrum of hues with materials that would have been quite familiar to Nicolas-Joseph Thiéry de Menonville. The techniques of a gifted contemporary artist and the biopiracy of an obsessed eighteenth-century botanist open windows onto the most enduring lessons that insects teach humans. Among these is the revelation that some of nature's smallest beings are bottomless reservoirs of possibility.

RESURGENCE AND RESILIENCE

It was a frosty February afternoon in 1828. Friedrich Wöhler could no longer contain his excitement. Breathless with elation over the mound of white crystals in the flask on his lab bench, the young German chemist wrote to his Swedish mentor, Jöns Jacob Berzelius: "I must tell you that I can prepare urea without requiring a kidney of an animal, either a man or dog." Wöhler had accidentally discovered how to synthesize the nitrogen-containing substance in mammal urine from two inorganic molecules, cyanic acid and ammonium. The Synthetic Age had begun.

Prior to the nineteenth century, the laboratory synthesis of natural products seemed unimaginable. This idea, rooted in concepts from ancient Greece, reached its apex in the nineteenth century. In 1810, John Wilkes's authoritative *Encyclopaedia Londinensis; or, Universal Dictionary of Arts, Sciences, and Literature,* a lavishly illustrated, twenty-four-volume compendium of Enlightenment knowledge, insisted that a "vital force" was present only in living organisms: "The substances which constitute the texture of vegetables differ from mineral substances in this, that they are of a more complex order of composition, and, though all are extremely susceptible of decomposition or analysis *not one is an object of synthesis.*" A mere eighteen years later, Wöhler's fabrication of a simple molecule shattered the time-honored distinction between organic compounds and inorganic substances.

During the twentieth century, the fascination with man-made

materials ranged far beyond the confines of laboratories. In 1923, the trendsetting Swiss-French architect Charles-Édouard Jeanneret, known to his contemporaries as Le Corbusier, announced, "The prime consequences of the industrial evolution in 'building' show themselves in this first stage; the replacing of natural materials by artificial ones, of heterogeneous and doubtful materials by homogeneous and artificial ones (tried and proved in the laboratory) and by products of fixed composition. Natural materials, which are infinitely variable in composition, must be replaced by fixed ones." Le Corbusier is widely remembered for his iconic pair of round, black spectacles and for his hard-edged functionalism, efficiently condensed into his maxim "A house is a machine for living in." This pragmatic style favored synthetic materials, the replacement of the craftsman with the engineer, and the standardization of design principles. Le Corbusier's influence was extensive. Clean-lined functionalist buildings that did away with ornate detail, prioritized function over form, and employed manmade substances were constructed in places as diverse as Switzerland, India, the Soviet Union, and Brazil.

The 1950s marked the high point of devotion to synthetic products and seemed to foretell the obsolescence of such naturally produced substances as shellac, silk, and cochineal. In 1952, *Science* published an article by the organic chemist Roger Adams called "Man's Synthetic Future." Adams, an industrious researcher who headed the University of Illinois Chemistry Department for nearly three decades, predicted the imminent demise of such everyday commodities as wool, cotton, silk, and leather. He boldly forecast their replacement with chemically synthesized compounds. As Adams put it, "In the future citizens will more effectively farm the land and the seas; obtain necessary minerals from the oceans; clothe themselves from coal and oil." Such transformations would come courtesy of modern chemistry.

Many of his contemporaries agreed. In their provocatively titled book *The Road to Abundance* (1953), chemist Jacob Rosin and writer Max Eastman claimed, "The time has come when the chemical industry can and will, slowly but surely, take over from agriculture the task of food production." A few years later, a special report prepared for the United States Senate Foreign Relations Committee concurred: "Synthetics may in the reasonably near future make the production of coffee, cocoa, cotton, sugar, wool and some other farm commodities unnecessary."

The conviction that humans possessed a boundless capacity to replace natural products with artificial substances surfaced in every corner of society. In 1957, the chemical company Monsanto collaborated with the Massachusetts Institute of Technology and the Walt Disney Company to create the "House of the Future," a prefabricated, cruciform dwelling constructed entirely of synthetic components. Over the course of a decade, more than twenty million visitors toured the 1,281-square-foot plastic and fiberglass home in Disneyland's Tomorrowland exhibition. Sleek, fully automated, and equipped with futuristic devices to irradiate food and sanitize kitchenware with sonic waves, the house prefigured life on *The Jetsons,* Hanna-Barbera's early 1960s animated sitcom about a utopian future. Predicting a world just over the horizon, the caption in a brochure that accompanied Disneyland's "House of the Future" confidently proclaimed, "In fact, it could be said that hardly a natural material occurs in its original state anywhere in

The "House of the Future," a model home made entirely from synthetic plastic and fiberglass. A collaboration among the chemical company Monsanto, the Massachusetts Institute of Technology, and the Walt Disney Company, the cruciform house was on exhibit at Disneyland from 1957 to 1967.

your new home!" Disney and Monsanto, inventors of chimerical and chemical futures, joined forces in this marriage of symbolic and literal elements. As philosopher Roland Barthes wittily described such collisions of myth and matter, "Despite having names of Greek shepherds (Polystyrene, Polyvinyl, Polyethylene), plastic . . . is in essence the stuff of alchemy."*

Although the Tomorrowland display promised visitors a window into "how the typical American family of four will live in ten years from now," the featured innovations were soon superseded by new ultramodern design trends. Disney dismantled the exhibit in 1967, but enthusiasm for synthetic products remained. That same year, the film *The Graduate* served up one of Hollywood's most enduring one-liners when Los Angeles businessman Mr. McGuire offered unsolicited career advice to a directionless college graduate named Benjamin, played by Dustin Hoffman: "I just want to say one word to you . . . plastics!" Synthetic materials offered salvation to wayward souls.

During the age of abundance that followed the Second World War, it seemed to many that the clock had run out for naturally produced insect commodities like shellac, silk, and cochineal. However, two factors intervened. The first was the rise of environmental toxicology. From the 1960s onward, practitioners of this newly constituted scientific discipline began to expose the lethal effects of synthetic chemicals released into the environment by design—such as pesticides and food additives—or by accident as industrial byproducts. These revelations demonstrated that humanity was marinating in a fatal brew of our own creation. As a result, consumer demand for alternatives stimulated a resurgence of naturally produced ingredients throughout much of the world. In many instances, manufacturers abandoned synthetic substitutes in favor of natural substances, including shellac and cochineal.

The second factor was the discovery that, despite enormous strides in polymer synthesis, industrial chemists still cannot manufacture inexpensive or structurally adequate alternatives to a surprisingly large assortment of naturally produced materials. Silk, like many other

---

* In 1907, Belgian chemist Leo Hendrik Baekeland (1863–1944) mixed phenol and formaldehyde under high heat and pressure to create a pliable material. This substance, known as Bakelite, was the world's first synthetic plastic.

natural products, has proven too complex to be efficiently engineered and profitably mass-produced using our existing knowledge and current methods. In an era dominated by the DuPont Chemical Company's 1935 promise of "Better Things for Better Living ... Through Chemistry," many organisms still operate as unrivaled laboratories of metamorphosis. As a result of surging demand for shellac, silk, and cochineal, people throughout the world continue to rely on the trade in insect secretions as a means of securing their livelihoods.

At precisely the same moment that synthetic futurism was mesmerizing the imaginations of millions, the applied science of toxicology tempered these lofty visions with a dose of reality. The study of the harmful effects of chemicals on living organisms has deep roots in antiquity. Initially, these investigations concerned the effects of natural, not synthetic, compounds on the human body. The Ebers Papyrus, a sixty-six-foot-long Egyptian scroll from the 1550s B.C.E., is the earliest known text to discuss the harmful properties of common substances. Full of esoteric incantations for repelling disease-causing demons, the manuscript also chronicles the adverse consequences for humans of exposure to materials such as hemlock and opium and the potential harmfulness of metals like lead and copper. More than a millennium later, medical practitioners in both ancient Greece and Han Dynasty China (206 B.C.E.–220 C.E.) relied on categories of substances—*pharmakon* and *du,* respectively—that could function as both poisons and antidotes, depending on their dosages and the methods of their application.

During the Middle Ages, scholars began to grasp the connections between occupational diseases and exposure to the hazardous materials miners encountered in their subterranean workplaces. The Swiss-German astronomer and alchemist Paracelsus (1493–1541) was among the first to express the foundational concept of "the dose makes the poison." From there he developed the bedrock ideas of modern toxicology. Such concepts took centuries to mature. It was not until the decades following the First World War that they coalesced as a self-contained discipline. The German *Archiv für Toxikologie* (Archive for Toxicology), first published in 1930, was the earliest scientific journal exclusively devoted to experimental toxicology. That same year, the U.S. Congress passed the Ransdell Act (Public Law 71-251), which for-

mally established the National Institutes of Health (NIH) and appropriated $750,000 (an enormous sum at the time) to fund new facilities
and sponsor research fellowships for undertaking basic biological and
medical research. A central concern for researchers at the new NIH
laboratories was the identification of toxic substances that U.S. citizens
encountered on a daily basis.

The decades following the Second World War were marked by a
cascade of environmental disasters and public health crises caused by
man-made chemicals. They included London's 1952 "Great Smog" of
airborne pollutants, the discovery in 1956 of widespread methylmercury poisonings among the population living near the Chisso Corporation's chemical factory in Minamata, Japan, and the worldwide tragedy
of the late 1950s and early 1960s, during which women in forty-six
countries who had used the sedative thalidomide during pregnancy
gave birth to some ten thousand physically deformed children. These
catastrophes, and others like them, stimulated the growth of toxicology into a full-fledged scientific field. As Julius M. Coon, the chair-
man of a 1960 meeting titled "Problems in Toxicology," announced to
a distinguished body of scientists assembled in a Chicago conference
room, "Toxicity is suddenly upon us as a social problem."

Two years later, Rachel Carson's groundbreaking book *Silent Spring*
exposed the detrimental effects synthetic pesticides were having
on human and environmental health. In unflinching prose, Carson
informed her readers about "500 new chemicals to which the bodies of
men and animals are required somehow to adapt each year, chemicals
totally outside the limits of the biological experience." Unsurprisingly,
some members of the chemical industry and some politicians reacted
with fierce opposition, attempting to discredit Carson on personal,
political, and scientific fronts. Former secretary of agriculture Ezra
Taft Benson gave these critiques their most crude expression. In a letter to Dwight Eisenhower, Benson questioned "Why a spinster with
no children was so concerned about genetics." He supplied his own
answer: Carson was "probably a Communist." Such unfounded criti-
cisms stung Carson. However, her editors and publicists had meticulously prepared for such an onslaught by vetting her book with experts
prior to publication and by arranging for television appearances that
bolstered her public persona. In the months following the book's

The marine biologist, writer, and conservationist Rachel Carson
(1907–1964) in Woods Hole, Massachusetts, in 1950. Carson's book
Silent Spring (1962) was a clarion call about the toxic effects of synthetic
pesticides on human and nonhuman beings. Most of her assertions
became widely accepted following the book's publication.

appearance, the attacks on Carson and *Silent Spring* lost their momen-
tum. The book sold six hundred thousand copies in its first year.

Revelations about the darker side of the Synthetic Age kept accu-
mulating. In the mid-1970s, jarring wake-up calls about the dangers
of chemical pollution included the disclosure that an entire planned
community in Love Canal, New York, sat atop twenty-one thousand
tons of toxic waste. As a report by an investigative subcommittee from
the U.S. House of Representatives remarked, "Industry has shown lax-
ity, not infrequently to the point of criminal negligence, in soiling the
land and adulterating the waters with its toxins." Of similar effect was
the news in December 1984 that nearly four thousand people had died
from a devastating leak of methyl isocyanate gas at the U.S.-owned
Union Carbide pesticide plant in Bhopal, India. (Government statistics now suggest that exposure to the gas caused fifteen thousand
deaths in the ensuing years.)

Pop culture kept pace with these developments. "Mercy Mercy Me
(The Ecology)," the 1971 hit by Motown icon Marvin Gaye, provided

a soulful anthem for a beleaguered planet. The end of its first stanza sets the tone:

> Where did all the blue skies go?
> Poison is the wind that blows from the north and south and
>   east.

By the late 1980s, four million synthetic chemicals were in production throughout the United States, and sixty thousand had entered into common use, leaching into the workplaces and homes of North America's cities and towns. Revelations about the human health consequences of this contaminated landscape began to receive widespread attention in the United States, Europe, and Japan. The historian Linda Nash has pointed out that "debates over chemicals and their regulation are, at root, debates about the relationship between bodies and their environments." A public discourse rapidly emerged about the toxic soup that was bubbling over in the midst of modern society.

The ensuing cultural backlash against the main tenets of the Synthetic Age was most apparent with the 1996 publication of *Our Stolen Future*. Featuring a foreword by Vice President Al Gore, the widely circulated book by a group of environmental health specialists asserted that synthetic compounds were drastically disrupting the functions of the human endocrine system, the hormone-secreting glands of the body. In the words of the historian Michelle Murphy, humanity had entered a new world order that amounted to a "chemical regime of living." The advent of this era also marked a shift in the timescale of environmental health issues. Hazardous substances that biodegrade over millennia leave a multigenerational imprint on both the ecosystems and the human tissues in which they accumulate. In 2018, medical researchers at the University of California, San Francisco, found an average of fifty-six potentially toxic chemicals in blood samples taken from seventy-five pregnant women delivering at two San Francisco hospitals.

On a finite planet, ideologies of limitless growth require corresponding theories of substitution. In an earlier era, colonies provided a temporary fix. By employing "ghost acreage," empires relied on spaces beyond their own homelands, such as oceans and occupied territories, to supplement their own harvests and limited resources. Yet in the cases of domesticated insects, their highly skilled cultivators, and their rarefied host plants, such dependencies often proved frustrating for imperial bureaucrats. Enduring misconceptions about the process of shellac production, failed attempts at sericulture, difficulties in successfully transferring the cochineal bug and its *Opuntia* cactus host to territories beyond Latin America, and the resistance of domesticated insect cultivation to economies of scale flustered Europeans in various ways. Often carried out with breathtaking ignorance of local knowledge, top-down attempts to turn nature into an assembly line have produced a litany of embarrassing failures and unmitigated disasters throughout the centuries.

The promoters of the Synthetic Age promised that the laboratory would provide a postcolonial escape hatch from such collisions with the limits to growth. Many of the synthetic substitutes for cochineal and shellac turned out to be toxic outcomes of the Synthetic Age. In 1950, children in the United States became violently ill from eating candy and popcorn colored with the man-made compounds FD&C Orange No. 1 and FD&C Red No. 32. New York congressman James J. Delaney subsequently organized hearings about chemical additives in the nation's food supply. The so-called Delaney Clause, which the U.S. Congress passed in 1958 as an amendment to the Food, Drug, and Cosmetic Act of 1938, stipulated that "no additive shall be deemed to be safe if it is found to induce cancer when ingested by man or animal, or if it is found, after tests which are appropriate for the evaluation of the safety of food additives, to induce cancer in man or animal." Naturally produced products, like shellac and cochineal, provided convenient, time-tested alternatives to these newly banned synthetic substances.

In addition to the health hazards of synthetic food additives, substituting artificial products for natural ones has proven intractable from the standpoint of both molecular chemistry and economic efficiency. Much like silk, substances as varied as blood, rubber, and vanilla are extremely resistant to replacement with manufactured alternatives. In the case of blood, the biggest obstacle to synthesizing artificial versions has been that oxygen-carrying hemoglobin becomes toxic outside the protective coating of the red blood cell. A 2019 article from *Popular Science* concluded, "At least for now, artificial blood remains a holy grail of trauma medicine."

Likewise, many of the most significant applications of latex from the Pará rubber tree (*Hevea brasiliensis*) cannot always be easily handled with synthetic latex. The tires of airplanes and colossal earth-moving vehicles must withstand tremendous pressures and high temperatures. Natural latex, sustainably tapped from trees as a milky liquid that drips from their bark, has sophisticated molecular properties and unprecedented pliability, features that meet these criteria. In addition, synthetic latex, developed during World War II as an alternative source of rubber during wartime shortages of this strategic material, produces allergic reactions in some consumers. Natural latex condoms, surgical gloves, and mattresses have all found emergent niches in the twenty-first-century market. Today, almost half of all rubber is derived from natural latex.

When it comes to vanilla, artificial and natural versions differ widely in purity, scent, and taste. A complex concoction of as many as 250 volatile aromatic compounds—along with a beguiling blend of chemicals that scientists refer to as nonvolatile tannins, polyphenols, resins, and free amino acids—augment the subtle flavor profiles of natural vanilla (almost exclusively *Vanilla planifolia*). Chemists working on synthetic vanilla have never been able to mimic this intricate amalgam, which contributes to the tastes and scents that appeal to many discriminating consumers and discerning chefs.

Silk has proved similarly inimitable. In a 2011 study, researchers at Oxford University and the University of Sheffield showed that "silk filaments produced by spiders and silk moths demonstrate combinations of strength and toughness that still outperform their synthetic counterparts." These scientists concluded that silkworms and silk-producing spiders had the crucial factor of time on their side: "With over 300 million more years of R&D it is not surprising that the animal's route to fiber formation is energetically optimized." The widespread recognition of such unique attributes has fostered a thriving market for silkworm fibers. The global trade in textiles made from *Bombyx mori* silk (known to economists as "mulberry silk") was $8.41 billion in 2016. The unit price for raw silk remains roughly twenty times that of raw cotton.

Contemporary sericulture shares many features with present-day shellac and cochineal production. The expansion of high-tech industries in East Asia has led many Japanese and Korean workers to abandon silkworm cultivation in favor of factory jobs producing circuit boards and cell phones. Meanwhile, villagers in China, India, Thailand, Vietnam, and Brazil have rapidly expanded their silk-raising ventures during the past few decades. Because sericulture is a home-based activity that responds well to microfinance, it provides opportunities for elevating the economic and social status of rural women. Although silk occupies only 0.2 percent of the total volume of the global textile trade, its commercial value is several orders of magnitude greater. Many nations produce primarily for internal consumption. For example, Indian silk weavers sell 85 percent of their production to domestic consumers, much of which ends up draped around women's shoulders and waists as sari cloths.

Sometimes silk saris save lives. Their tightly woven mesh features pores so small that these garments provide effective drinking-water filtration. When silk or cotton saris have been folded back upon themselves multiple times, they can be placed over a collecting urn to strain out phytoplankton and microorganisms in river and canal water. As microbiologists have confirmed, these sari filters successfully capture *Vibrio cholerae,* the microscopic comma-shaped bacteria that cause the life-threatening disease cholera.

While in some settings silk is an everyday fabric with life-saving potential, in others it continues to serve as a provocative status symbol. Allegra Versace, the grown daughter of Italian fashion designer Donatella Versace and former U.S. fashion model Paul Beck, told *Harper's Bazaar* in 2007, "My mom dressed me in silk to go to elementary school. . . . In kindergarten, they sent me home because I couldn't do finger painting in my dress."

Silk has taken the beauty industry by storm in areas other than fabric. Skin-care companies in Asia have embraced the manifold marketing possibilities for silkworm cocoons. About the size and color of a cotton ball, these compressed bundles of thread are natural storehouses of the protein sericin, which helps prevent dehydration and boosts the moisturizing properties of skin creams and salves. Products developed in Thailand and South Korea, like "cocoon facial soap" and "cocoon peeling silk balls," have stimulated global beauty trends.

The revitalization of silk because of its unique properties has much in common with the factors that have driven the resurgence of shellac and cochineal. Furniture restorers and music connoisseurs have spearheaded the rediscovery of shellac. As *New Yorker* writer and home-improvement guru David Owen has commented, "Shellac's main appeal as a wood finish is aesthetic. It enhances the natural color and grain of many varieties of wood, and it doesn't dry to the dead, filmy sheen of polyurethane." The revival of William Zinsser's French polishing technique has ensured shellac's place on the workbenches of both do-it-yourself hobbyists and master artisans. By applying many thin layers of shellac with a rag or brush, carpenters and musical instrument makers can create a high-gloss surface that enhances the fine grains and the rich tones of woods like mahogany and rosewood.

These connoisseurs share a fondness for shellac with a budding group of enthusiasts who feel nostalgic for a golden age of recording. American music critic Amanda Petrusich spent several years getting to know the aficionados who obsessively collect shellac 78 rpm records. As she concluded, there is "something profoundly tactile and enigmatic about a 78 and the sounds it contained." Shellac also turns up in some less expected venues. Dedicated anglers who hunch over lamps and hover over magnifying glasses, carefully tying "flies" for trout, salmon, and bass fly-fishing, often use shellac to waterproof their intricate feather, fur, thread, and hook lures. In a curious interplay of emulation and dependency, fly-fishers use an insect product (two, if the thread is silk, as it so often is) to mimic a caddis fly, a nymph, or a flying ant with which to entice their quarry.

These examples seem to suggest that shellac is merely a relic, resurrected by antiques dealers, hard-core hobbyists, and high-end niche markets. Yet shellac has emerged as a key material in the next generation of food preservatives, pharmaceuticals, cosmetics, and electronics. Because of its approval by the United States Food and Drug Administration (FDA) and the European Union, shellac is routinely used by food manufacturers as an ingredient stabilizer, a food thickener, a produce coating, and a packaging adhesive.

Halloween trick-or-treat bags, grocery store fruit bins, and household medicine cabinets are also full of shellac-coated products. Confectionary companies employ shellac as a safe glaze to prevent the chocolate and colorings in candies from melting and running. Sweet treats, from malted milk balls to jelly beans, feature this glossy, nontoxic veneer. Fruit growers spray shellac on apples to make them shine and coat oranges, lemons, and avocados with this insect product to extend their shelf lives. Likewise, the pharmaceutical industry relies on shellac as an enteric agent. A shellac coating slows the release of the active ingredients in pills and tablets when medicines are exposed to the human stomach's harsh acidic environment. Cosmetic producers use shellac (often labeled as "gum lac") in aerosol sprays, lotions, shampoo, nail polish, lipstick, eyeliner, and mascara. The bodies of the dead and the mouths of the living are also full of shellac. When preserving cadavers, embalmers now often rely on this insect compound as a nontoxic substitute for formaldehyde. Meanwhile, dentists routinely use shellac-based composites in fluoride varnishes, dentures, and fillings.

In a more unexpected plot twist in the continuing saga of shellac, this insect secretion is also playing a pivotal role in the electronics industry. A rapidly expanding area of the technology sector is the development of green, organic equipment. The magnitude of disposed electronic devices (known as e-waste) generated globally each year is staggering, reaching fifty million tons in 2018. As Achim Steiner, executive director of the United Nations Environment Programme (UNEP), told journalists, this amounts to a "tsunami of e-waste rolling out over the world." The need for alternatives is urgent. Researchers in Austria, Romania, and Turkey have used shellac (along with such organic substances as indigo, paper, silk, gelatin, and vegetable starches) as a key component in the manufacture of organic field-effect transistors, devices that will help make electronics biodegradable and allow for their safe use in and on human bodies. The reduced cost, low toxicity, and minimal environmental impact of such microelectronics will not only decrease the long-term waste streams, but may also open new avenues for therapeutic and health-monitoring applications.

Although Cambodia, Indonesia, Myanmar, Thailand, Vietnam, and China all produce shellac for international markets, most of the world's shellac still comes from India, which exports $45 million annually. Today, the eastern Indian state of Jharkhand—literally "the land of the forest"—is often called "the lac state." This rural region is responsible for more than half of the country's shellac output. In Jharkhand and elsewhere, women have been central to the revitalization of lac production. Given their systematically disadvantaged position in the labor market and their limited access to capital, poor women in outlying regions often depend on the collection of non-timber forest products (NTFPs), such as lac. Lac rearing involves as many as five million people in India, and NTFPs continue to provide a primary source of income for up to 75 percent of rural women across the nation.

Much the same can be said of cochineal. Its production remains a thriving cottage industry, dominated by entrepreneurial women and rural families. Today, Peru, Bolivia, Argentina, the Canary Islands, and Botswana have joined Mexico as the major exporters of this high-value insect dye. The global market for cochineal coloring reached $16.7 million in 2017 and has grown by nearly 6 percent annually since 2014.

In an era when producers are seeking nontoxic colorants with long shelf lives and durability across a wide range of conditions, cochineal has, once again, become a much-sought-after dye. Buried deep within the ingredients lists that grace the labels of everyday products, it masquerades under many pseudonyms. Its immense array of applications, listed earlier, results from cochineal's exceptional physical properties. It is among only a few water-soluble colorants that resist degradation over time. The unusual stability of cochineal during cooking and freezing and its resilience in acidic environments are noteworthy. Yet despite these remarkable qualities, a thorough understanding of cochineal's chemistry was slow to emerge. Although a German chemist had isolated the compound responsible for the insect's red pigment in 1818, it was not until 1959 that scientists fully understood the structure of carminic acid. This research was put on the back burner as synthetic dyes rose to prominence in the post–World War II era.

The decades that followed the war exhibited a mixture of optimism and anxiety. The Synthetic Age coincided with the Atomic Age. These interrelated upheavals in human history marked the dawn of a malleable world in which humans tinker with the fundamental processes that govern life and death.

Insects played a special role in this era of creative hubris and existential angst. For over half a century, a somewhat misleading notion has circulated that cockroaches can survive a nuclear holocaust. This belief began with a jumble of anecdotes suggesting that these insects had withstood the 1945 U.S. atomic bombings of Hiroshima and Nagasaki, devastating explosions that delivered fatal levels of radiation to the human inhabitants of both cities.

Such claims about cockroach hardiness in the wake of nuclear fallout received backing from improbable sources. In 1962, the Johns Hopkins University geneticist H. Bentley Glass enlivened the myth of the postapocalyptic cockroach in a speech he delivered at Smith College. Glass provocatively asserted that in the aftermath of a nuclear war, "the cockroach, a venerable and hardy species, will take over the habitations of the foolish humans and compete only with other insects or bacteria." *The New York Times, The Nation,* and several other major U.S. publications reprinted Glass's remarks, effectively making his comments "go viral" in a pre-Internet era.

Three years later, amid increasing revelations about U.S. entanglements in Vietnam, the National Committee for a Sane Nuclear Policy, an ad hoc organization dedicated to promoting debate on the hazards of nuclear testing, ran a full-page advertisement in *The New York Times.* The provocative illustration, reminiscent of Albrecht Dürer's stag beetle from 1505, featured a plump cockroach set against a stark white backdrop. The caption read, "The winner of World War III," while the text at the bottom of the page declared, "A nuclear war, if it comes, will not be won by the Americans . . . the Russians . . . the Chinese. The winner of World War III will be the cockroach." Such assertions left little doubt in readers' minds about the resilience of these insects, even in the face of the era's most pressing threat to human existence.

In the pages of the March 9, 1970, issue of *New York* magazine, the journalist Catherine Breslin put these features into stark relief: "For sheer survival the [cockroach] is remarkably equipped. . . . Step on him and he'll compress his outer skeleton; lift your foot and he'll whistle into a hairline crevice. Freeze him alive and he'll amble off when thawed. Starve him out and he'll live two months on water alone. When hard pressed he'll swim, fly, digest wood, eat his own cast-off skin or somebody's wife's eggs." These hyperbolic claims about cockroach hardiness persist into the twenty-first century. Pixar's 2008 film, *WALL-E,* depicts a postapocalyptic Earth bereft of humans and barren of other life. The planet's lonely waste-disposal robot finds companionship in a cockroach named Hal.

Although fables of the indestructible cockroach are captivating, the science does not support them. As physicists Claus Grupen and Mark Rodgers point out, "The idea that cockroaches have an extremely high radiation tolerance is a myth: although higher than that of humans, their resistance is similar to that of many invertebrates." During the 2008 season of the Discovery Channel television series *MythBusters,* the cast of hip do-it-yourself scientists exposed German cockroaches to 10,000 radon units (rads) of cobalt 60, which promptly exterminated all of the bugs. The atomic bomb the *Enola Gay* dropped on Hiroshima emitted gamma rays at a strength of around 10,000 rads.

Even so, cockroaches are one of the planet's toughest customers. They are "living fossils," retaining a physical structure that has remained relatively unchanged for an astounding 340 million years. They are also everywhere. The American entomologist Samuel Hubbard Scudder, arguably the nineteenth century's leading authority on cockroaches, remarked, "Of no other type of insect can it be said that it occurs at every horizon where insects have been found in any numbers."

To those who do not share Scudder's unbridled enthusiasm for the members of this insect family, such pervasiveness is a chilling thought. In the popular imagination, cockroaches are disease-ridden scavengers that scurry out from behind our refrigerators at night and haunt the undersides of our trash cans by day. What is not common knowledge is that these all-too-familiar domestic companions represent less than half a percent of the nearly five thousand described species of cockroaches on the planet. The vast range of cockroach sizes is one of the clearest testaments to their astounding diversity. During their rare aboveground appearances, the males of Australia's wingless burrowing cockroach (*Macropanesthia rhinoceros*) mature to three inches in length and are easily mistaken for small tortoises. At the other end of the spectrum are the mosquito-sized *Attaphila fungicola* roaches, which loiter among the fungal gardens tended by some of their larger insect cousins, the leaf-cutter ants of the tropical Americas. When these tiny tenants fancy a departure from the confines of their host colony, they stealthily hop aboard a departing winged ant and ride it to a new destination.

This penchant for covert mobility spans nearly all cockroach species. Cockroaches are extraordinarily adept hitchhikers. In *A Generall History of Virginia, New England, and the Summer Isles* (1624), Captain John Smith mentioned "a certaine India Bug, called by the Spaniards a Cacarootch . . . creeping into Chests they eat and defile with their ill-sented dung." When paired with Smith's gloomy testimonial about the Virginia Colony's failed attempts to raise silkworms, his dispirited account suggests that the English soldier and explorer had not experienced the most favorable interactions with insects.

The Middle Passage also expedited the spread of cockroaches to the Americas. The *Wanderer,* one of the last ships to bring an illegal cargo of slaves from Africa to the United States, was a floating habitat for far more beings than slave traders and their prisoners. As journalist Erik Calonius notes, "The cockroaches that had crawled aboard in Africa had multiplied. Now, not a corner of the ship was without them." For the hundreds of kidnapped children, women, and men who endured this dreadful six-week crossing from the mouth of the Congo River to the tidal marshlands of Jekyll Island, Georgia, cockroaches surely intensified the filth and terror of their captivity.

Other cockroach species have also followed distinctive routes in their global dispersion. In 1886, two British zoologists explained that by the 1500s, the Asian cockroach (*Blatta orientalis*) "appears to have got access to England and Holland, and has gradually spread thence to every part of the world." Cockroaches made an effortless transition from the age of oceanic shipping to an era of global airborne transit. On July 18, 1989, Chinese authorities discovered 13,262 German cockroaches (*Blattella germanica*) aboard seventeen aircraft that landed at Guangzhou Baiyun International Airport, one of China's busiest travel hubs.

Cockroaches have also crossed the border between fact and fiction. Indeed, generations of readers have assumed that a cockroach is the metamorphic outcome of Gregor Samsa's night terrors in Franz Kafka's celebrated *The Metamorphosis*. Gregor, who wakes one morning to find himself inexplicably transformed into a monstrous insect, spends the remainder of his time reflecting on his deprived existence. However, as the translator Susan Bernofsky notes in her 2014 rendition of the novella, "In Kafka's correspondence with his publisher, he was adamant that the 'insect' (*Insekt*) not be depicted on the jacket of the book. And although he and his friends used the word 'bug' (*Wanze*) when referring casually to the story, the language that appears in the novella itself is carefully chosen to avoid specificity." Gregor may have been a cockroach or a distant insect relative.

The year after the publication of *The Metamorphosis*, New York *Evening Sun* columnist Don Marquis gave a more lighthearted twist to his six-legged protagonist, introducing the memorable character Archy the cockroach. Archy, a free-verse poet reincarnated in insect form, taps out satirical poems and social commentary on a creaky old typewriter. Because Archy jumps from letter to letter to type, he almost never manages to operate the shift key (too much of a stretch for those tiny legs). As a result, most of his stanzas appear in lowercase letters. Over the years, hundreds of Archy's witty musings reached readers through Marquis's daily column, "The Sun Dial."

Archy's jovial disposition contrasts with the pitiful circumstances of the central character in the traditional Spanish folk song "La Cucaracha" ("The Cockroach"). The origins of this tragicomic ballad about an insect that struggles to walk *"porque le faltan las dos patitas de atrás"* ("because it is missing the two hind legs") are shrouded in mystery, but there is ample evidence that the lyrics changed as the tune traveled

A kinder, gentler cockroach and his feline friend, Mehitabel.
In 1916, columnist Don Marquis created "Archy," an affable poet,
reincarnated in insect form. For more than a decade, this prolific bug
shared his musings with readers of the New York *Evening Sun*.

from the Iberian Peninsula to the Americas. "La Cucaracha" acquired a decidedly political tenor during the Mexican Revolution when the cockroach represented the vilified figure of Victoriano Huerta, a Mexican military officer who coordinated the overthrow and assassination of Mexico's beloved president Francisco Madero in 1913. Francisco "Pancho" Villa and Emiliano Zapata led the rebel armies that ousted Huerta the following year. "La Cucaracha" served as a sardonic anthem for their troops.

Nearly a century later and half a world away, the implications of the term "cockroach" reached new lows, becoming directly associated with genocide. During the spring of 1994, as extremists from the Hutu majority in Rwanda turned on their Tutsi neighbors, *inyenzi,* or "cockroach," became the epithet of choice for the Hutu hard-liners who orchestrated the mass murder of hundreds of thousands of Tutsis. In her haunting memoir *Cockroaches,* the author Scholastique Mukasonga describes her childhood as a Tutsi girl in the Rwandan countryside, her departure for Burundi the year before the horrors began, and the slaughter of her family in the genocide that engulfed the small African nation.

Shockingly, such atrocious similes retain their vigor. On April 17, 2015, the far-right media personality Katie Hopkins wrote a column in Britain's tabloid *The Sun* about North African refugees fleeing their devastated homelands for Europe's shores. "These migrants are like cockroaches," she declared. "They might look a bit like 'Bob Geldof's Ethiopia circa 1984,' but they are built to survive a nuclear bomb." Hopkins penned her hateful screed just forty-eight hours before a mass drowning of seven hundred migrants after their boat sank sixty miles off of the Libyan coast.

Genocidal rhetoric, cautionary tales, and humorous escapades aside, the cockroach exhibits a steadfast tenacity in quite literal (and nonliterary) ways. In a 2019 study, entomologists at Purdue University reported that widely dispersed German cockroaches, the same peripatetic species whose members clambered aboard airplanes bound for China, are rapidly outwitting human attempts to regulate their reproductive lives. As the researchers discovered, German cockroaches can develop immunity to new insecticides within a single generation, making roach control with man-made chemicals nearly impossible. Although these creatures may not be able to endure a nuclear apocalypse, cockroaches are finding new ways to survive the chemical regimes of the Synthetic Age.

Discoveries about cockroaches and their kin are even toppling conventional systems of organizing the world. In 2007, a team of scientists used genetic analyses to determine that termites, often incorrectly referred to as "white ants," are actually a family (Termitidae) of cockroaches. These newly inducted members of the cockroach clan share the evolutionary fortitude of their relatives.

Their resilience is underscored in Stephen Vincent Benét's 1933 poem "Metropolitan Nightmare." This prophetic tale of urban life deteriorating under the stresses of a warming world opens with foreboding: "That was the year the termites came to New York / And they don't do well in cold climates." By the final stanza, we learn that the termites have adapted more successfully than their human neighbors. As "an old watchman, beside the first / Raw girders of the new Planetopolis Building," tells a young reporter:

> "Oh, they've quit eating wood," he said, in a casual voice,
> "I thought everybody knew that."
>   —and, reaching down,
> He pried from the insect jaws the bright crumb of steel.

Although Benét was trafficking in fantasy, he was remarkably prescient about real-world climate change and the creatures most adept at handling such unprecedented environmental stresses. Insects inhabited this planet long before us, they have adapted in resilient ways to our disruptive presence, and they are likely to inherit what humans leave behind.

This hardiness is no guarantee that the crucial, but often inconspicuous, roles insects play as decomposers, pollinators, and food sources in countless ecosystems will remain undamaged. Insects, like so many other organisms, are suffering mightily as part of a broader global decline in biodiversity known as the sixth extinction: the sixth time in the planet's history when vast numbers of species have vanished with stunning rapidity. This time, however, it is humans, not ice ages or asteroids, causing these mass disappearances.

The first global scientific review of insect extinction rates, published in 2019, concludes that upwards of 40 percent of insect species are in decline and a third are endangered. The latest data reveal that the total mass of insects on Earth is currently falling by 2.5 percent each year, a terrifying statistic; it suggests the total vanishing of most insects within a century. As the journalist J. B. MacKinnon puts it, "Extinction is not mere death; it is the death of the cycle of life and death." Given the vital tasks insects perform to ensure our planet's cycles of reproduction and decay, MacKinnon's summation could hardly be more suitable for any other group of organisms.

The wholesale reconsideration of the assumptions that animated the Synthetic Age has helped insects and the products they create experience a full-blown resurgence. Shellac, silk, and cochineal have made surprising comebacks in areas that even the most adept forecasters could hardly have predicted. The next step in this course correction will be to fundamentally alter the practices—pesticide- and herbicide-dependent agriculture, carbon-intensive energy systems, and habitat-destroying development strategies—that threaten the extinction of our six-legged cousins. Nature exhibits an astonishing degree of resilience, but this capacity to recover and adapt is not infinite. As the cockroach has shown us, even the hardiest of creatures will succumb to extreme conditions.

PART II

HIVES OF MODERNITY

6

NOBEL FLIES

A century ago, modern genetic science was born in a cluttered office on Manhattan's Upper West Side. Room 613 of Columbia University's Schermerhorn Hall was barely the size of a studio apartment and reeked of rotting bananas and stale tobacco smoke. The ten oak desks crammed into the tiny laboratory were strewn with milk bottles salvaged from the alley behind the university cafeteria. Inside this hodgepodge of glass containers, thousands of flies, devouring morsels of overripe fruit, swarmed like bargain hunters at a clearance sale. On any given day from 1911 to 1928, a dozen young men and women huddled over their brass microscopes, puffing at pipes and cigarettes while searching for mutations among each new generation of their winged subjects. These researchers affectionately referred to their cluttered lab as the "Fly Room."

This nickname was apt. The anesthetized insect beneath each microscope lens was a common fruit fly.* Featuring bulging vermilion eyes and a translucent khaki-colored body, *Drosophila melanogaster* matures to only about one-third of the size of a housefly (*Musca domestica*). *Drosophila*'s deep-seated association with dumpsters and landfills belies its enormous importance to the history of modern science. This tiny creature's genome—its complete set of DNA, includ-

---

* The term "fruit fly," adopted in the early twentieth century, often causes confusion. It can also refer to Mediterranean fruit flies, known to scientists as *Ceratitis capitata*.

ing all of its genes, which constitutes the blueprint for its biological development—has served as the Rosetta Stone for cracking the code of human genetics.

When most of us picture the typical laboratory animal, mice and rats scurry to center stage. However, the *Drosophila* fly is the most widely used model organism in biomedical research. For more than a century, this fruit-basket busybody has been central to many of the key discoveries related to the biological mechanisms that govern animal development and disease responses. As of 2018, eight Nobel Prizes have been awarded to eighteen scientists for research involving common fruit flies. Thomas Hunt Morgan, the energetic founder of Columbia's Fly Room, won the first of these prestigious honors for his team's *Drosophila* research. The pioneering discoveries he and his protégés made about the role of the chromosome in heredity offered a road map for the rapidly developing field of genetics.

As a child, Morgan was aware of his family's reputation for greatness. His ancestors included John Wesley Hunt, one of the first millionaires west of the Allegheny Mountains, and Francis Scott Key, the composer of "The Star-Spangled Banner." He was born in Lexington, Kentucky, on September 25, 1866, five weeks after President Andrew Johnson declared an official end to the War Between the States. Morgan's parents, Charlton Hunt Morgan and Ellen Key Howard Morgan, were influential southern planters who had fervently supported the Confederacy. Prior to the U.S. Civil War, Charlton had served as U.S. consul to Messina, Italy. He became the first diplomat to recognize the government of Giuseppe Garibaldi, the charismatic nationalist general who led many of the military campaigns that unified Italy. Ironically, Garibaldi was a committed supporter of the Union and even offered his military services to Abraham Lincoln.

During the war, Charlton served as a Confederate aide at Manassas (the First Battle of Bull Run) and was later wounded and apprehended by Union soldiers at the Battle of Shiloh. After his release, he volunteered in the Kentucky regiment commanded by his brother. Charlton accompanied John until United States troops captured them in 1863.

Family history shaped Thomas Morgan's curricular choices at the State College of Kentucky, forerunner to the University of Kentucky. He enjoyed studying French, but a vindictive language professor nearly gave him a failing grade. The instructor had fought for the Union in the Civil War and been captured by "Morgan's Raiders," a cavalry unit organized by Morgan's great-uncle, the Confederate general John Hunt Morgan, that conducted raids deep behind Union lines. General Morgan had forced the young soldier to ride a mule backward for ninety miles from Cincinnati to Lexington.

With his linguistic ambitions thwarted, Thomas Morgan gravitated toward the natural sciences, which anchored his successful undergraduate career. He graduated as valedictorian of the class of 1886. Morgan went on to pursue his doctorate in zoology at Johns Hopkins University. From 1888 onward, he spent his summers at the newly opened Marine Biological Laboratory (MBL) in the seaside town of Woods Hole, Massachusetts. A former whaling village perched on the southwestern corner of Cape Cod, this hub of biological research fostered Morgan's innovative approaches to experimental science and piqued his curiosity about ocean-dwelling organisms. He took full advantage

Thomas Hunt Morgan (1866–1945) at work with his microscope in Woods Hole, Massachusetts. Conducting hundreds of thousands of experiments on the common fruit fly (*Drosophila melanogaster*), Morgan and his colleagues built the foundations of modern human genetic research.

of the freshly collected marine specimens the MBL offered to its scientists, writing his doctoral dissertation on the embryology of sea spiders, bottom-feeding marine arthropods with tiny bodies and long, spindly legs that inhabit most of the world's oceans.

Morgan's sea spider research landed him his first job. In 1891, he became associate professor of biology at the recently founded Bryn Mawr College for Women, one of the first institutions in the United States to offer postgraduate education to women. The following summer, Morgan returned to Woods Hole, where zoology professor Edmund Beecher Wilson introduced him to his future biology graduate student Lilian Vaughan Sampson. Romance found inspiration at the lab bench. Morgan and Sampson fell in love and eventually married, in 1904.

Sampson, who took Morgan's surname, became a major geneticist. In the 1920s, her groundbreaking research on *Drosophila* significantly advanced our understanding of chromosomes, literally "colored bodies," the threadlike structures in most living cells that transmit an organism's genetic information. Remarkably, Lilian published sixteen single-authored papers in major scientific journals while raising a family and managing her husband's financial affairs for five decades.

In 1904, Thomas Morgan accepted a position at Columbia University as professor of experimental zoology. He and Lilian moved into a house on West 117th Street in Manhattan, where they had four children. The Fly Room was a ten-minute walk from their home, across the tree-lined paths and grassy fields of Morningside Park. Morgan remained at Columbia until 1928, when he was tapped to head a new school of biology at the California Institute of Technology, in Pasadena, where he continued his fruit fly research. That year, at a reception in his honor, he mused about the preordained nature of his westward migration: "Of course I expected to go to California when I died, but the call to come to the Institute arrived a few years earlier, and I took advantage of the opportunity to see what my future life would be like." Morgan was not the first exile from the East Coast to compare California to heaven.

Morgan won the Nobel Prize in Physiology or Medicine in 1933 for his fly-based discoveries about the role the chromosome plays in heredity. Not given to pomp or spectacle, he chose to skip the lavish presentation ceremony in Stockholm. After some cajoling from his colleagues, including a case of Prohibition whiskey provided by

Caltech's board of trustees, he agreed to travel to Sweden the following year to claim the award. In April 1934, Morgan, Lilian, and one of their grown children took passage to Europe aboard the steamship *Majestic*. On their way, they stayed overnight in New York with old family friends. The sixty-seven-year-old Morgan "appeared at the door," his host recounted, "dressed typically in a rather disreputable old topcoat. In one pocket he had a comb, razor, and toothbrush wrapped in newspaper; in the other he carried a similarly wrapped pair of socks. 'But what else do you need?'" asked the bemused Morgan.

During his later years, Morgan often conducted research at the Kerckhoff Marine Laboratory in Corona del Mar, California. This villa, clad in white stucco and roofed with red tiles, featured a running seawater system and wet experiment tables designed to replicate the facilities of the world-renowned Stazione Zoologica in Naples, Italy, where Morgan had spent ten months in the mid-1890s. He stayed at Caltech until his death, in 1945, from a ruptured artery.

Throughout their itinerant life, he and Lilian remained devoted to their summertime community of Woods Hole. For nearly half a century, the couple made their annual pilgrimage to Cape Cod. Morgan served as an MBL trustee from 1897 to 1945, helping to manage the affairs of an institution that rapidly matured into one of the world's premier centers for research and education in biological and environmental science. The Morgan family maintained a welcoming wood-shingled summer cottage near the beach that frequently hosted as many as two dozen relatives and graduate students.

Like the Morgan household, the town of Woods Hole was a hive of activity, sustained by the comings and goings of collaborators and scientific colleagues from across the globe.* This invigorating intellectual atmosphere was matched by the ecological diversity and visual splendor of the region's renowned estuaries, beaches, and salt marshes. Rachel Carson recalled, "I had my first prolonged contact with the sea

---

* Incidentally, I owe an enormous debt of gratitude to the Woods Hole waterfront and Lilian Morgan for my own revelations about the natural world. In 1913, Lilian cofounded the Summer School Club at Woods Hole to further her commitment to outdoor science instruction for youngsters. The organization later became the Children's School of Science. For many summers, this gabled two-story clapboard house nestled on a hillside above Eel Pond Harbor was the site of my own epiphanies about wild things, including my affinities for monarch butterflies and other six-legged creatures.

at Woods Hole. I never tired of watching the tidal currents pouring through the hole—that wonderful place of whirlpools and eddies and swiftly racing water."

Despite the allure of the Cape Cod seashore, Lilian and Thomas made their most profound contributions to science through insect research conducted in the more austere confines of their Manhattan Fly Room. Morgan's legacy is forever linked to the "white-eyed mutant male" *Drosophila* fly he discovered in April 1910. This curious creature was a striking anomaly among its red-eyed cohort. Morgan had bred thousands of fruit flies with crimson-colored eyes. With the appearance of this white-eyed ocular oddball, he had finally found what he had sought for so many years: an abnormal inheritable trait.

Morgan diligently cared for this peculiar fly. He kept it in a tiny flask that he tucked into his coat pocket at the end of each workday. At the time, Lilian had just given birth to the couple's third child, a girl. Such was the elation surrounding Thomas's discovery that, on his first visit to the hospital, Lilian asked, "How's the fly?" Thomas replied, "How's the baby?"

While Lilian and her daughter—also named Lilian—adjusted to a postpartum world, Thomas hunkered down to begin experiments with his unusual insect. He bred this male mutant with several red-eyed female flies. With only 3 exceptions, all of the 1,237 offspring had red eyes. Next, he mated the red-eyed females with his original white-eyed mutant. Previous research on inheritance suggested that the next generation would feature a ratio of one white-eyed fly to every three red-eyed flies, regardless of sex. Although the eye colors among this new generation did display these expected proportions, the inheritance pattern was unequal among males and females. The vast majority of the white-eyed offspring were male, just like their extraordinary grandfather. By showing that inheritance was linked to the male or female sex of the offspring, Morgan blazed a new trail. *Science* published his revolutionary findings, under the title "Sex-Limited Inheritance in *Drosophila,*" in July 1910. Morgan subsequently linked the inheritance of specific traits with particular chromosomes. Over the course of his prolific career, he published twenty-two books and 370 scientific papers, which laid the foundations for modern genetic science.

Morgan chose a minuscule fruit fly for his research because of the unusual attributes of this creature. In Groucho Marx's droll phrasing, "Time flies like an arrow. Fruit flies like bananas." There is truth in comedy. *Drosophila* flies thrive on little more than a buffet of mashed overripe fruit served in a moist environment. It is no coincidence that the Greek words for "dew-loving" are the source of *Drosophila*'s name. The life of the common fruit fly is a Dionysian frenzy of sugar and sex. In the words of the historian Robert Kohler, *Drosophila* is "a biological breeder reactor, creating more material for new breeding experiments than was consumed in the process." At room temperature, a single mated pair of flies can produce five hundred eggs in two weeks.

Their swan song arrives rapidly. Offspring attain sexual maturity within a week and live for only about thirty days. Crucially, each *Drosophila* fly possesses four very large chromosomes that are easily visible under a light microscope. Another scientifically useful feature of the common fruit fly is the magnitude of its nervous system. As the science writer Jonathan Weiner put it, "An *E. coli* bacterium is a single cell . . . a nervous system with a single neuron. At birth, a human baby has about one hundred billion neurons, one for every star in the Milky Way. A fruit fly has about one hundred thousand neurons, so it is the geometric mean between the simplest and the most complicated nervous system we know." Morgan and his colleagues in the Fly Room took full advantage of these favorable attributes.

Scientific discoveries rarely occur in a social vacuum. Most depend upon intricate networks of connections among collaborators, both

Because of the rapidity with which it reproduces and the simplicity of its physiology, the common fruit fly (*Drosophila melanogaster*) has been among the most popular model organisms in the history of science.

past and present. True to form, the Fly Room's researchers drew inspiration from the recently revived findings of Gregor Mendel, an obscure Augustinian monk. During the warmer months of the mid-1800s, a bespectacled man in a black cassock could be found tending to the meandering green tendrils of several dozen edible pea plants (*Pisum sativum*) in the five-acre garden of St. Thomas's Abbey in Brünn, Austria. The city, now called Brno in Czech, became part of the Czech Republic after the First World War. It has long been famous for the imposing baroque fortress of Špilberk Castle, which became the Austro-Hungarian Empire's notoriously brutal prison for political dissidents. Brno is also the gateway to the celebrated vineyards of South Moravia, home to the aromatic white-grape wine known in German as *Welschriesling*. Today, one of the principal destinations for Brno's visitors is a tidy bed of red and white flowers nestled behind a wrought-iron fence. The chessboard pattern of blossoms in the monastery's garden pays homage to Mendel's now-famous experiments on plant heredity.

For millennia, farmers have crossbred various strains, breeds, or varieties to achieve desirable traits in plants and animals. The diversity on display in seed catalogs, dog shows, and cattle auctions attests to these hard-won legacies. Sweet corn, Meyer lemons, Irish wolfhounds, and Hereford-Angus cows are well-known exemplars of interspecies hybridizing. As the son of peasant farmers, Mendel had practiced such age-old strategies in his family's fruit orchard. He later deduced the mechanisms behind the making of hybrid vigor, discovering that genes come in pairs and get passed on, one from each parent, as distinct units. Mendel developed his ideas of dominant and recessive traits to describe the segregation of parental genes and their expression in offspring. As he discovered, the logic of inheritance could be described with modest mathematical formulas. These revelations would eventually transform the time-honored art of breeding into an established scientific practice.

Mendel's laws of inheritance, which would profoundly influence Thomas Hunt Morgan's conclusions about fruit flies, germinated during seven years of elegant experiments on thirty-four subspecies of pea plants. Over eight growing seasons between 1856 and 1863, the friar fastidiously counted at least forty thousand blossoms and three hundred thousand peas that grew in his garden and an adjacent greenhouse. Like Morgan, Mendel was tenaciously chasing variety. Plant height (tall or short), seed color (green or yellow), seed shape (smooth or wrinkled), seed-coat color (gray or white), pod shape (full or constricted), pod color (green or yellow), and flower distribution (along the stem or at the end of the stem) were the seven isolated traits Mendel scrutinized when cross-pollinating pea varieties. After breeding a new generation of plants, he could then note which visible features were passed on to the new offspring.

Mendel's contemporaries subscribed to the older theory of blended inheritance, the notion that parental traits were mixed in offspring and could not be separated back into their original forms. As this dominant school of thought suggested, all of these traits would eventually produce a diluted amalgamation of the parental characteristics. Mendel's garden peas told a different story, one that would become the first chapter of a longer saga about genetics that Morgan and his colleagues would help write.

On a clear, crisp February evening in 1865, a stocky middle-aged man, "friendly of countenance, with a high forehead and piercing blue eyes, wearing a tall hat, a long black coat, and trousers tucked into topboots," strode into an imposing building known as the Modern School and gave the first of two lectures he would deliver to the Natural History Society of Brünn. While describing his startling results to the three dozen scientists in attendance, Mendel introduced two new laws of heredity. The first of these was the law of segregation: even though an organism inherits two "factors" (which scientists now call genes) from its parents, it contributes only one of them to its offspring. The second was the law of independent assortment: traits are transmitted to offspring independently of one another. Mendel explained these principles in a paper, "Experiments on Plant Hybrids," which appeared in the society's journal the following year.

Although these astonishing statements should have shaken up the scientific world, they languished in obscurity for the remainder of the century.* After Mendel became his monastery's abbot in 1867, he found

---

* Mendel's legacy has not been without its controversies. In 1936, the British statistician Ronald Fisher argued that the Moravian monk had fudged his evidence to more precisely fit his assumptions about the predicted ratios of inherited plant traits. In 2010, two Portuguese statisticians urged moderation, concluding that Mendel's data were more likely the product of unconscious biases than the result of outright fraud.

himself consumed by administrative duties, including a protracted dispute with government officials over the taxation of religious institutions. He made few efforts to publicize his work beyond distributing forty reprints of his article. One of Charles Darwin's students later discovered Mendel's paper—its bound pages never cut open—squirreled away among the 1,480 titles in the crowded library at Darwin's country estate outside London. Lamentably, the greatest naturalist of the nineteenth century apparently ignored the remarkable contents of Mendel's publication. Several decades after Mendel's death, from a kidney disorder, at age sixty-one, a friend recalled, "Not a soul believed his experiments were anything more than a pastime, and his theories anything more than the maunderings of a harmless putterer." Mendel's anonymity, however, was short-lived.

In the early 1900s, the British biologist William Bateson, a mustachioed, outspoken don at St. John's College, Cambridge, became a fierce champion of the Moravian friar's findings. Bateson, the first scientist to describe the study of heredity as "genetics," expounded on the earth-shattering implications of Mendel's research: "Investigations which before had only been imagined as desirable now became easy to pursue, and questions as to the genetic inter-relations and compositions of varieties can now be definitely answered." Bateson's timing was opportune. Other European botanists were simultaneously rediscovering Mendel's discoveries and replicating his groundbreaking results. U.S. researchers, including Thomas Hunt Morgan and his team of fly geneticists, followed their transatlantic counterparts and joined the Mendelian renaissance.

Just as Morgan had initially resisted ideas of chromosomal inheritance, he never embraced all the basic tenets of Mendelian heredity. By drilling down to the level of the chromosome and the gene, the scientists in the Fly Room challenged some of these premises. Such rejections of established scientific wisdom typified Morgan's career. From his graduate school days onward, he developed a reputation for flouting convention. As his long-term collaborator, the geneticist Alfred Henry Sturtevant, recalled, this iconoclasm typified Morgan's experiments:

> For example, if you mix eggs and sperm from the same individual, normally nothing happens. But sometimes self-fertilization

does occur. And one of the questions was, Why? What brings this about? How does it happen? And Morgan had a nice hypothesis: maybe the acidity of the water is responsible. Let's see what pH changes will do. But being Morgan, he didn't set up measured amounts or concentrations. What he did was to take a dish in which eggs and sperm were present and squeeze a lemon over it. And it worked. Then he studied it in more detail after that. This was one of the most successful experiments in the field.

Morgan's propensity for aggressive experimental research rewarded him in 1915. That year, he and his Fly Room collaborators, Alfred Sturtevant, Calvin Bridges, and Hermann Joseph Muller, unified their extensive laboratory findings in a book they titled *The Mechanism of Mendelian Heredity*. This landmark text refined Mendel's laws and combined them with the theory, first suggested in 1902 by the German biologist Theodor Boveri and the American geneticist Walter Sutton, that chromosomes are the bearers of hereditary information.

At first, Bateson overtly resisted this revision of Mendel's theories, broadcasting his strong objections in the scientific press. It was only after a weeklong visit to "the *Drosophila* workers," as he called Morgan and his collaborators, in their home environment of Columbia University's Fly Room, that Bateson acquiesced to Morgan's groundbreaking fusion of Mendelian inheritance and chromosomal theory. Even in defeat, Bateson favored the dramatic flourish. In the winter of 1921, he announced to his colleagues at the Toronto meeting of the American Association for the Advancement of Science: "I come at this Christmas season to lay my respectful homage before the stars that have arisen in the west." A multigenerational collaboration among a humble Moravian monk, a Yankee from the heart of the Confederacy, a prolific patch of garden peas, and generations of fruit flies thus laid the cornerstones of classical genetics.

Such a motley assortment of human and nonhuman characters would never have converged to unravel the mysteries of genetic inheritance without the alignment of many fortuitous circumstances. Among these was the nineteenth-century migration of the common fruit fly to the United States. *Drosophila melanogaster* is a forest-dwelling species native to equatorial Africa. At some point—still hotly debated by evolutionary biologists—it made the dietary leap from eating wild plants to dining on domesticated fruits. As an organism that lives with another species without harming its host, *Drosophila* is what ecologists call a "commensal." This term carries an intriguing double meaning. Its original connotation was "one who eats at the same table." Indeed, *Drosophila* has entwined its alimentary preferences with ours. Among the first reports of this gastronomic liaison was an 1864 discovery of a swarm of *Drosophila* flies feasting on mounds of sultana raisins in an Ottoman storehouse. A decade later, the first North American account of *Drosophila* came when entomologist Joseph Albert Lintner remarked that these tiny flies had been "bred from a jar of pickled plums" he had received in the mail. From this traveling container of conserves, common fruit flies radiated outward, eventually finding happy homes in supermarket aisles and kitchen fruit baskets across the United States. As Alfred Sturtevant later recalled, the produce section of a Woods Hole grocery store was the birthplace of Morgan's first generation of fruit flies.

Like the silkworm, the cochineal bug, and the shellac insect, *Drosophila* hangs its hat wherever its human hosts dwell. The common fruit fly is not only a commensal, it is what biologists refer to as a "cosmopolitan organism." From the narrow confines of its evolutionary homeland in the jungles of equatorial Africa, it has now traversed the farthest latitudes and has been spotted as far north as Finland and as far south as Tasmania.

Drosophila flies are also getaway artists, managing Houdini-like escapes from laboratories, kitchens, airplanes, and cargo ships. In the early 1940s, they scattered across O'ahu, Hawai'i, after bolting from culture bottles and zooming out the windows of a university laboratory. Two decades later, fruit flies took wing from the food locker at the Charles Darwin Research Station in the Galápagos Islands. This isolated volcanic archipelago where the twenty-six-year-old British naturalist developed his theory of evolution by natural selection in 1835 was now home to a new triumph of adaptation.

During the interwar years, *Drosophila* became a common model organism in the ivory tower. In 1937, the Irish marine biologist Ernest William MacBride remarked upon the groundswell of fruit fly–based genetic science that surged across Europe and North America: "Everywhere [academic] chairs in genetics have been established, nay more, in some places special chairs in 'Drosophily' devoted to the exclusive study of this type-animal." As the twentieth century unfolded, university laboratories set out the welcome mat and left the lights on for the common fruit fly.

Drosophila's ascendant public persona developed in tandem with the more secretive, even clannish, aspects of fruit fly experimentation. As though they are hawking contraband, *Drosophila* researchers refer to their craft as "fly pushing." Although one of the promises of modernity was the replacement of superstition with rationality, the language of magic permeates genetic studies. The leading *Drosophila* research manual observes, "At times, it may even appear that the wielders of 'hard-core' fly genetics preside over a coven with secret rites of initiation." Likewise, a 2007 genetics research article asks if a particular type of molecule that regulates gene expression in *Drosophila* is "the magic wand to enter the Chamber of Secrets," and a 2014 paper expounds on the "Wizardry and Artistry of *Drosophila* Genomics." Apparitions of fantasy and myth in the midst of the "hard science" of genetics surely stem from the fact that a little winged bug has been so central to deciphering the mysterious codes that govern human development.

Indeed, the boundary between science and magic is not as clear-cut as so many Science 101 textbooks maintain. Think of flight, telescopic vision, space exploration, or molecular engineering. Such technological breakthroughs were once the dreamy aspirations of the now-forgotten arts of alchemy and astrology.

Insects also span the chasm between tradition and modernity. Arrayed in a brittle carapace, antennae twitching atop a pair of bulbous compound eyes, they hybridize our dinosaur daydreams with alien iconography, simultaneously reminding us of aeons past and foreshadowing what life-forms may dwell beyond our interplanetary horizons.

Yet beneath these dislocations in time and space lies a steadfast familiarity between these creatures and us. This unwavering intimacy manifests itself in the Romantic English poet William Blake's classic 1794 poem "The Fly":

Am not I
A fly like thee?

Or art not thou
A man like me?

More recently, Sarah Lindsay's 2002 poem "The Common Fruit Fly"
imagines the insect's viewpoint on the long arc of human research with
Drosophila. The first two stanzas establish a tone of lament:

Five hundred generations
since we traced the sweetness above a windfall apple.
Instead we taste your scheduled anesthesias
and wake on a shelf in flasks to be filled
with our stubborn procreation.
You pour us out on Mondays.
You sort us with thick fingers.

We are the white-eyed, the red-eyed, the eyeless,
with full wings, with crumpled vestigial wings.
We perceive the buzz of fluorescence you call quiet,
its flicker you call light.
The chemical stream of your midafternoon banana
sings behind our mouthparts.
We are the tested.

"The tested" ultimately provided crucial models for research projects with implications that stretched far beyond the confines of the Fly
Room. Morgan's protégés used these exemplary insects to make a stunning number of genetic discoveries that extended well into the latter
half of the twentieth century. Hermann Muller zapped *Drosophila* to
prove that X-rays caused lethal genetic mutations, while Alfred Sturtevant used the flies to demonstrate linkage, the tendency of genes that
are located in close proximity to be inherited together. Breakthroughs
about the workings of cancer cells, immunity to certain fungal and
bacterial infections, and the intricate functioning of neural networks
have all been made through fruit fly research.

Experiments done on *Drosophila* have even offered compelling
insights about how sleep cycles function in animals, including humans.
Three American researchers—Jeffrey C. Hall, Michael Rosbash, and

Michael W. Young—won the 2017 Nobel Prize in Physiology or Medicine for demonstrating that when a single "period" gene in the fruit fly undergoes a mutation, it disrupts the insect's circadian rhythm. Scrutinizing the workings of this "inner clock," which regulates sleep-wake cycles, is a rapidly expanding branch of biology. Circadian rhythms also modify feeding behavior, hormone release rates, blood pressure, and body temperature in humans, making this area of study crucial to the health and well-being of our species.

At times, the importance of fruit fly research has been lost in the political spin cycle. During the months before the 2008 U.S. election, vice presidential candidate and Alaska governor Sarah Palin heaped scorn upon *Drosophila* research as an example of wasteful government spending. At a campaign rally in Pittsburgh, she attempted to rile up her audience by denouncing the fact that "[your tax] dollars go to projects that have little or nothing to do with the public good—things like fruit fly research in Paris, France. I kid you not." Palin's ignorance of the inestimable scientific value of fruit fly experimentation was especially ironic, given that her youngest son, Trig, suffers from Down syndrome, a genetic disorder that has been extensively studied by scientists using *Drosophila* chromosomes.

Today's media landscape, driven by sound bites and one-liners, has made it easy to dismiss insects as frivolous distractions. However, as the story of Thomas Hunt Morgan's breakthroughs elegantly demonstrates, a white-eyed mutant fly no bigger than an asterisk was the key to unlocking the mysteries of inheritance. In the immortal words of America's foremost poet, Walt Whitman, "The nearest gnat is an explanation." The unity of the universe is revealed in all things.

LORDS OF THE FLORAL

I t was a quiet winter day in January 1862 when a parcel arrived at the front door of Down House, a Georgian manor fifteen miles south of London. The package had begun its journey two hundred miles to the northwest at Biddulph Grange, a Staffordshire estate once described by *The Independent* newspaper as "one of the most extraordinary gardens in Britain. . . . It contains whole continents, including China and Ancient Egypt—not to mention Italian terraces and a Scottish glen." James Bateman, the man who mailed the box, was a master orchid grower who had sponsored several plant-collecting expeditions to the tropics.

When the parcel's recipient, Charles Darwin, cracked open the wooden cover of Bateman's packing case and dug beneath a layer of straw, he discovered a meticulously arranged assortment of rare flower specimens. Among the precious cargo was a milky white, star-shaped orchid with a foot-long nectar tube. Darwin was astounded by this anatomical oddity. He immediately dashed off a letter to a close friend and colleague, the English scientist Joseph Dalton Hooker, exclaiming, "Good Heavens what insect can suck it?" The remarkable flower that stirred Darwin's imagination was the Madagascar star orchid (*Angraecum sesquipedale*). A few days later, Darwin sent a second note to Hooker in which he made one of evolutionary biology's most audacious predictions: "In Madagascar there must be moths with proboscides capable of extension to a length of between ten and eleven inches."

Thomas William Wood's illustration of Charles Darwin's hypothetical
pollinator for the Madagascar star orchid, published in the October 1867 edition
of the *Quarterly Journal of Science.* Darwin's prediction proved remarkably
accurate, and the moth's eventual discovery verified the naturalist's claims
about the relationship between an insect pollinator's physiology
and the anatomy of its host flower.

Later that year, Darwin published his hypothesis about the coevolution of a plant and its pollinator.* Many of his contemporaries reacted with disdain. The creationist philosopher George Campbell devoted several pages of *The Reign of Law* (1867) to mocking the "long noses" of Darwin's stealthy moths. Campbell claimed that among Darwin's conclusions "we find nothing but the vaguest and most unsatisfactory conjectures."

An exception to the skeptics was the Welsh-English naturalist Alfred Russel Wallace. The frequently overlooked codiscoverer of natural selection stepped forward to champion Darwin's prediction. Wallace described a plausible anatomy for Darwin's mystery creature, basing his conception on a West African species of moth known to

---

* When two (or more) species repeatedly alter each other's traits over time, scientists consider them to be coevolving.

scientists since the 1830s. He commissioned the talented zoological illustrator Thomas William Wood to sketch an image of his proposed moth. Wood produced a drawing akin to a wanted poster for the mysterious pollinator. The illustration featured an imaginary tangle of lush tropical flowers and jungle vines. In Wallace's opinion, the existence of such an insect was assured: "That such a moth exists in Madagascar may be safely predicted; and naturalists who visit the island should search for it with as much confidence as astronomers searched for the planet Neptune, and I venture to predict they will be equally successful!" Wallace was not given to modesty.

In 1903, Darwin and Wallace were triumphantly vindicated. That year, the zoologist Walter Rothschild and the entomologist Karl Jordan published the first description of the sphinx moth *Xanthopan morganii praedicta,* "the predicted one." Rothschild and Jordan publicly confirmed that Darwin and Wallace had correctly anticipated the extraordinary tongue of the Madagascar star orchid's sole pollinator. Scientists later discovered that this hummingbird-sized moth emerges under cover of darkness, sniffs its floral quarry to ensure that it has targeted the correct variety of blossom, and unreels a wiry proboscis (between eight and fourteen inches long), which extends like a taught fishing line into the nectar pool of the orchid's deep receptacle. In a catalog of Earth's most exquisite adaptations, this plant-insect duo would surely be a prized entry.

While siphoning the puddle of sweet juice at the bottom of this tube, the sphinx moth receives a generous dusting of the orchid's pollen grains. As it brushes along the petals, the moth's scaly body becomes an airborne conduit, transporting these valuable grains to other Madagascar star orchids and pollinating them. Even though a few photographs of this highly choreographed ritual surfaced in the 1990s, it was not until 2004 that video footage emerged. That year, Professor Philip DeVries, a tropical butterfly and moth expert from the University of New Orleans, hiked into a Madagascar jungle in the dead of night, positioned an infrared camera next to a star orchid, and managed to obtain an astonishing movie of the sphinx moth inserting its prodigious appendage into the orchid's nectary. Like a clip of surveillance footage from a security camera, the grainy black-and-white movie captures the exchange. The moth hovers in front of the milky orchid, uncoils its feeding tube, and lingers for a drink. Then it flits off to its next refueling station, carrying with it the granules that will ensure the continuity of the orchid's reproductive cycle.

In various guises, and with a wildly diverse array of contrivances, this fleeting act of aerial transmission plays out trillions of times each day around our planet. Pollination is the most vital stage of a seed plant's life cycle. For fertilization to occur, pollen grains containing the plant's male sex cells must travel aloft from the stalk-shaped stamens that produce them (or the male cones of a pine) to the ovule-bearing organs—the pistils of flowering plants (or the female cones of coniferous plants). A successful transfer of pollen from stamen to pistil triggers seed production. Some angiosperms (as flowering plants are known in scientific parlance) rely on gusts of wind to broadcast their pollen, while over 90 percent depend on the Velcro-like surfaces of the fur, feathers, or scales of animal carriers. Most of these pollinators are insects. They are, to borrow Darwin's words, "Lords of the floral." Pollination is rarely a one-sided bargain. As compensation for their roles as pollen carriers, insects acquire nectar, obtain shelter, and gain access to rich storehouses of the chemical constituents of pheromones, the chemicals insects use to communicate with each other.*

For many of us, the powdery yellow haze of spring initiates a frenzy of seasonal rituals at car washes and drugstores. Submerged in the deluge from watering eyes and runny noses, we can forget that the orgy of plant sex in the air around us is vital to our well-being. This coevolutionary dance of powder and pollinators is responsible for one in every three bites of food the average human takes. Annually, upwards of half a trillion dollars' worth of crops grown around the world—including avocados, lemons, almonds, blueberries, eggplants, watermelons, coffee, and tea—relies on the highly specialized activities of pollinators.

Most of the ambassadors in this vital interchange are bees. Some twenty thousand species of wild bees—plus many species of butterflies, flies, moths, wasps, and beetles—constitute an entomological federa-

---

* Although interactions between plants and pollinators are generally reciprocal—plants gain the transfer of pollen among flowers and pollinators receive rewards like nectar—a spectrum of relations exists. Some visitors to the floral bouquet are true-blue thieves, while certain plants lure guests without offering compensation for their services.

tion of pollen transmitters. Their liaisons with the botanical world are coordinated by a flower's fragrance, shape, size, and color palette, features that serve as declarations of plant fertility. Some plants have even developed the ability to change their floral colors as a means of signaling which flowers are most fecund and therefore ready to dispense or receive pollen. The tiny, bright blossoms of the tropical lantana shrub (*Lantana camara*) publicize their first day of blooming by turning a rich golden hue. At this stage they are lush with nectar and pollen. On the days that follow, these gilded petals blush, turning orange and scarlet to signal that pollinators should pay more attention to neighboring flowers on the bush.

To many human spectators, insect pollination suggests spontaneous meandering. The poet Mary Oliver tenderly referred to this seemingly impromptu insect undertaking as "stopping / here and there to fuzzle the damp throats / of flowers." According to recent research, however, there is far more intention to the pollinator's itinerary than meets the human eye. Scientists in England have used radar tracking devices to chart bee flight pathways during foraging expeditions. They have discovered that bees are master navigators, amassing cartographic wisdom from experience. As the bees repeat their food searches, they generate more efficient routes, rearranging the sequence of floral stopovers and altering their flight paths between nectar sources.

Two pioneering scientists, working on opposite sides of the Atlantic, made fundamental discoveries about the purposeful and perceptive ways that bees interact with flowers and intermingle with their fellow hive mates. American Charles Henry Turner (1867–1923) and Austrian Karl von Frisch (1886–1982) achieved their entomological breakthroughs in less than ideal circumstances. Turner was born just two years after the Civil War to a church custodian and a nurse. Despite facing racial discrimination throughout his scientific career, he became the first African American to receive a PhD in zoology from the University of Chicago and the first African American ever to be published in the prestigious journal *Science*. Von Frisch, who went on to win a Nobel Prize in Physiology or Medicine in 1973, was targeted by the Nazi regime for his Jewish heritage and his practice of employing Jewish research assistants, many of whom were women.

While Turner made multiple contributions to the study of insects, among his most significant was his demonstration that members of the Hymenoptera—a large order of insects, comprising the sawflies, wasps, bees, and ants—are not simply reflexive machines, as so many of his contemporaries thought, but are instead organisms with the capacities to remember, learn, and feel.

Among the seventy-one papers Turner published during his distinguished thirty-three-year career, two stand out for their conclusive demonstrations that honeybees can perceive colors and patterns. To test his hypotheses, he designed thirty-two simple but elegant experiments, which he performed under the pin oaks, magnolias, and silver maples that ring the scenic lake at O'Fallon Park in St. Louis. In one of these tests, Turner placed a dish filled with jam atop a picnic table every morning, noon, and evening for several days. The bees visited the sweet buffet three times each day. Turner then stopped putting out the jam at lunch and dinnertime and set out a dish only at breakfast. For a few days, the bees continued to appear at all three times, but they soon adjusted their behavior, visiting the table only in the morning. Turner determined that bees have a sense of time and an ability to rapidly develop new feeding habits in response to changing conditions.

Another of Turner's innovative honeybee experiments established the role of bee eyesight in guiding flower pollination. During the early 1900s, biologists were aware that, at close range, flowers attracted bee pollinators by producing certain scents, but these researchers knew next to nothing about the visual components of such attractions when bees were too far from the flowers to smell them. To explore this topic, Turner pounded several rows of wooden dowels into the park's lawn. Atop each rod, he attached red disks dipped in honey. In short order, bees began to visit his makeshift flowers for their sugary treats. He then added a series of sticks topped with blue disks and no honey. The bees ignored the new "flowers." Turner then drizzled honey on the blue disks, and the bees gradually caught on, visiting these as well. Turner concluded that the bees had deduced an initial connection between red and the presence of honey. Color served as a visual signal when the bees were farther away from Turner's flowers, but they were able to smell the honey on the blue disks at closer range and adapt.

In just over three decades of research, Turner carried out an astounding range of experiments. The conclusions from these wide-

Charles Henry Turner (1867–1923) was the first African American
to receive a PhD in zoology from the University of Chicago and the first
African American to be published in the prestigious journal *Science*.
His innovative field experiments proved that honeybees possess advanced
cognition and use eyesight to guide their pollination choices.

ranging tests helped establish his reputation as an authority on the
behavioral patterns of bees, cockroaches, spiders, and ants. In 1910,
the French naturalist Victor Cornetz honored his North American
colleague by naming the exploratory circling movements that foraging ants make when returning to their nests *tournoiement de Turner*
("Turner circling").

Despite being more prolific than many of his white contemporaries
who held distinguished appointments at major research universities,
Turner was unable to secure a long-term college or university position.
The University of Chicago would not offer him a job, and Booker T.
Washington could not afford to hire him at the Tuskegee Normal and
Industrial Institute. After a stint as an instructor at the University of
Cincinnati and a brief period as a biology professor at Clark College
(now Clark Atlanta University), Turner spent the remainder of his
career teaching at Sumner High School in St. Louis. His starting salary
in 1908 was $1,080 a year (around $30,000 in today's dollars).

Sumner was the earliest high school for African Americans established west of the Mississippi River. Its prestigious faculty included Edward A. Bouchet, whose 1876 physics doctorate from Yale made him the first African American to receive a Ph.D. from a U.S. university. At Sumner, Turner made his discoveries about insect behavior without access to fully equipped laboratory facilities, a research library, or graduate students. Despite these challenges, he left a lasting legacy. Subsequent studies by entomologists have verified most of Turner's assertions about insect behavior in general and honeybee pollination in particular.

Turner was only fifty-five when he died of acute myocarditis, an infectious heart inflammation. He passed away just four years before the publication of a landmark study that would dramatically advance human understandings of bees. Karl von Frisch's 1927 book *Aus dem Leben der Bienen* (*The Dancing Bees*) deciphered the "round dance" and the "waggle dance" of the Carniolan honeybee (*Apis mellifera carnica*). This subspecies of western honeybee, native to central Europe, is favored among beekeepers for its pugnacious temperament toward insect pests and its tranquillity around humans. As von Frisch discovered from decades of patient observation and experimentation, the two dances performed by this preferred pollinator are elaborate but precise ways of sharing news about food sources near the hive.

The round dance is a flamboyant demonstration by a foraging bee to its hive mates that there is a feeding site between fifty and a hundred yards away from the hive. Von Frisch wrote a poetic description of the Sufi-like twirls of ecstasy that typify this communication ritual:

> The foraging bee, having got rid of her load, begins to perform a kind of "round dance." On the part of the comb where she is sitting she starts whirling around in a narrow circle, constantly changing her direction, turning now right, now left, dancing clockwise and anti-clockwise in quick succession, describing between one and two circles in each direction. This dance is performed among the thickest bustle of the hive. What makes it so particularly striking and attractive is the way it infects the surrounding bees; those sitting next to the dancer start tripping after her, always trying to keep their outstretched feelers in close

contact with the tip of her abdomen. They take part in each of her maneuverings so that the dancer herself, in her madly wheeling movements, appears to carry behind her a perpetual comet's tail of bees. In this way they keep whirling round and round, sometimes for a few seconds, sometimes for as long as half a minute, or even a full minute, before the dancer suddenly stops, breaking loose from her followers to disgorge a second or even a third droplet of honey while settling on one, or two other parts of the comb, each time concluding with a similar dance. This done, she hurries towards the entrance hole again to take off for her particular feeding-place, from where she is sure to bring back another load; the same performance being enacted at each subsequent return.

As von Frisch painstakingly determined, the bees he was studying also performed a second form of communicative ritual, the "waggle dance," to relay information about more distant food sources. Using the vertically hanging honeycomb of the hive as her stage, the dancing bee shimmies forward a certain distance. She then traces a half circle back to her starting point and begins the dance again. During the straight stretch of the dance, she "waggles" her rear end, revealing the direction of the food source to her hive mates. The angles of her undulations indicate the relationship between the flight trajectory and the position of the sun, while the time she takes to traverse the straight stretch of her intricate ballet conveys the distance to the feeding spot. Von Frisch discovered that these dance-based itineraries were highly successful guides to food, even if the bees had to travel past mountains or over forests to reach their destinations.

Writing in the journal *Nature,* one of the early reviewers of *The Dancing Bees* maintained that von Frisch was "the most distinguished of living experimental zoologists, and it is to him that we are indebted for the greater part of recent advance in knowledge of the behavior and sense organs of the honey bee." The astonishing conclusions von Frisch made about honeybee communication met with some opposition at first, but by the end of the twentieth century his fundamental theories had been substantiated.

Such accolades did not come without considerable suffering. Von Frisch was born in Vienna on November 20, 1886, the same year Sig-

Karl von Frisch (1886–1982), the Austrian zoologist who discovered "the language of bees." During the 1930s, he was targeted by the Nazi regime for his Jewish heritage. He endured many years of harassment, but he went on to win the Nobel Prize in 1973.

mund Freud, another Jewish intellectual hounded by the genocidal Nazi regime, opened his first psychiatric office in the city. Von Frisch studied medicine and zoology in his hometown and in Munich. After teaching at several other universities, he was appointed chair of the Zoology Department at the University of Munich, where he used funds from a $372,000 Rockefeller Foundation grant to establish a new institute in his field of animal studies.

The same year the three-level, state-of-the-art Institute for Zoological Research opened in Munich, a succession of shocking political events reached their crescendo about three hundred miles to the north, in Berlin. Following a series of parliamentary elections and backroom political intrigues, the German president Paul von Hindenburg appointed Adolf Hitler as chancellor of Germany on January 30, 1933. A few months later, the newly empowered National Socialist regime passed the Law for the Restoration of the Professional Civil Service, which required that German civil servants furnish "proof of Aryan ancestry" (*Ariernachweis*). This menacing directive prompted an immediate exodus by many prominent Jewish and leftist intellectuals. Albert Einstein promptly resigned from the Prussian Academy of Sciences and immigrated to the United States. Von Frisch remained at his teaching post in Munich but was unable to provide a certificate of ancestry for his maternal grandmother, and he acquiesced to officials who claimed that she was of non-Aryan descent.

Initially, the university rector allowed von Frisch to retain his professorship under the label "quarter Jew." Swept up in a wave of "denunciation fervor," militantly anti-Semitic students contested this decision and campaigned for his dismissal. Wilhelm Führer, the vehemently anti-Semitic leader of the Munich Instructors League, submitted a formal testimonial that described von Frisch as "a petty-minded, narrow specialist who has no understanding of the new era and is extremely hostile toward it." Among the damning assertions in Führer's letter was his claim that von Frisch had shown "unusually great partiality for Jews and Jewish associates by marriage," building a laboratory around values that were decisively opposed to the racist goals of the Nazi *Weltanschauung*, or worldview.

Von Frisch persevered through years of harassment and continued to teach and conduct research on bees. He survived the war and emerged after the Nazi defeat with his career intact. Much like Turner, who was obstinate in the face of overt discrimination, von Frisch seems to have been driven by a deep commitment to continuing his research through any hardship. As one of von Frisch's biographers put it, some heady questions were at stake in his work: "Were nonhuman animals capable of symbolic communication? And if so, what did this mean for a scientific understanding of the animal-human boundary?"

The modern-day business of migratory beekeeping in the United States puts into stark relief the fragility of this boundary and humans' continued dependence on insects. Mobile apiculture is an industry of massive proportions. It involves the long-distance transport of trillions of bees around the country to pollinate food crops. California's Central Valley, which enjoys three hundred days of sunshine each year and a twenty-five-degree temperature fluctuation from day to night (an ideal hot-cold cycle for many food crops), is one of the world's most fertile agricultural zones. Like a gigantic hothouse, the eighteen-thousand-square-mile region (approximately the area of the Dominican Republic) produces more than half of the nuts, fruits, and vegetables grown in the United States. Citrus fruits, plums, avocados, melons, cherries, apples, strawberries, blueberries, lettuce, squash, broccoli, and almonds are the most prominent pollinated crops in the state's cornucopia. The quality of the harvest and the quantity of the yield depend on the industrious undertakings of insect pollinators from beyond state lines.

Among these cultivated plants, almonds have one of the most compelling stories about our enduring reliance on insect pollinators. Seemingly endless rows of stocky nut trees that bear clusters of pink-and-white flowers dominate vast expanses of California's agro-industrial landscape. The state's massive almond (*Prunus dulcis*) crop spans a 1.5-million-acre area that stretches from Sacramento to Los Angeles and features more than thirty varieties of one of the world's most popular nuts. Once these oblong seeds have been removed from their hard shells, the almonds end up salted and stuffed into airline snack packs, pulverized and diluted to make almond milk, slivered into salads, ground into almond butter, and drenched in layers of chocolate and caramel.

None of these American almond products would be possible without the legions of migratory beekeepers who make their living by transporting their insects from overwintering sites in places like Florida to farms across the country. Each year, between October and February, thousands of tractor trailers shuttle hives housing thirty-one billion European honeybees (*Apis mellifera*) across the nation's highways to California. The insects from these mobile apiaries pollinate the state's capacious kingdom of nut trees. As the logic goes, since the trees cannot move, the pollinators must come to them. This is far too mammoth a task for wild pollinators, so human "bee brokers" match hives of domesticated European honeybees with orchards to choreograph the annual ritual of plant sex that keeps more than seven hundred billion almonds popping off the trees every year. This crop is big business: almonds are the seventh largest U.S. agricultural export.

The insects these migratory beekeepers use to perform their vital service are not native to North America.* Scientists have long warned of the risks that come with relying solely on exotic honeybees for crop

---

* Much like their modern-day pollinators, almond trees are not native to North America. Almonds were first domesticated in Persia and came to California with the Spanish during the eighteenth century.

pollination. There are more than four thousand species of wild bees native to North America. These include bumblebees, mason bees, orchid bees, and many other solitary species in the family Apidae that, on a bee-to-bee basis, are more efficient pollinators of certain New World plants than imported honeybees, which are native to Eurasia. For example, some North American bumblebees have evolved to be able to "unlock" a flower's pollen supply by vibrating their abdomens.

Regardless of whether the insects are native species or more recent arrivals to the environments of their preferred flowers, it is hard to imagine a planet without pollination. This tried-and-true mechanism of reproduction, relied upon by the majority of the world's flowering plants, has been around for at least one hundred million years. Minuscule powder-covered bugs sequestered in golden chunks of amber from northern Spain offer mesmerizing fossilized windows onto the insect's ancient role in plant fertilization.

Many generations prior to European contact, civilizations on several continents demonstrated a sophisticated awareness of the roles pollinator insects played in maintaining the health of the plant world. The Yolngu people of northern Australia developed totemic relationships with insects, thereby safeguarding bees and their habitats. Nahuatl and Totonaco farmers from the Sierra Norte of Puebla, Mexico, established oral traditions forbidding the killing of bee and butterfly pollinators. And the Seneca people, one of the five nations of the Haudenosaunee (Iroquois) Confederacy, amassed generations of detailed knowledge about the winged creatures that sustained communities of edible wild flora and medicinal herbs. Centuries of formal research and multifaceted traditions of indigenous ecological knowledge have affirmed that the successful coupling of insects and flowers is vital to the health of our biosphere.

The now-famous maxim, widely attributed to Albert Einstein, "If the bee disappears from the surface of the earth, man would have no more than four years to live" was not, in fact, his. Instead, it had far more obscure origins in a 1941 article by columnist Ernest A. Fortin published in the *Canadian Bee Journal*. Nevertheless, this declaration of dependency conveys much wisdom.

To answer the question of what a world without bees would look like, the American conservation biologist Thor Hanson suggests a visit to the Juan Fernández Islands. These rugged volcanic outcroppings, located four hundred miles off the coast of Chile, were the temporary home of Alexander Selkirk, a Scottish sailor who spent four years and four months marooned there, beginning in 1704. Selkirk's story later served as the inspiration for the Englishman Daniel Defoe's novel *Robinson Crusoe*. Bereft of human company and short on provisions, Selkirk subsisted on a makeshift diet of spiny lobsters, feral goats, white turnips, and cabbage until his rescue by the English privateer Woodes Rogers. During Selkirk's prolonged stay on the craggy isles, the castaway must have noticed the scarcity of floral life. The sporadic clusters of drab greenish-white flowers on this beeless, rocky archipelago have adjusted their pollination strategies to the whimsy of the winds and the comings and goings of birds. According to Hanson, the only type of bee to ever make the transoceanic journey to the Juan Fernández Islands "is a tiny, rare sweat bee thought to have arrived recently from coastal Chile." This insect émigré is not a significant pollinator for the islands' plant life.

In recent years, credible fears of an insect apocalypse have inspired renewed visions of a hardscrabble life in a barren world without these crucial pollinators. During the winter of 2006–07, beekeepers in the United States and Europe began reporting widespread deaths among their colonies of European honeybees. As many as half of all affected colonies displayed symptoms that did not match the typical patterns of normal honeybee die-offs. The media were quick to pick up on early warning signs of an impending disaster. A March 2007 headline in *The New York Times* announced, "Beekeepers Confronted by Demise of Colonies," while *The Guardian* followed up the next month with a story titled "Threat to Agriculture as Mystery Killer Wipes Out Honeybee Hives." Before long, scientists had begun referring to such crises as Colony Collapse Disorder (CCD).

Not all aspects of this bee catastrophe were without precedent. In the United States, the first recorded case of such a phenomenon occurred more than 150 years ago. In 1868, the U.S. commissioner of agriculture reported, "During the past season a disease suddenly appeared in Indiana, Kentucky, and Tennessee, sweeping away whole apiaries. So quiet were its operations that the bee-keepers became aware of its existence only by the disappearance of their bees. The hives were left, in most cases, full of honey, but with no brood and little pollen; the whole appearance of the hive causing the casual observer to suppose that the bees had 'emigrated'; but close observation showed that they had died." The cause of this die-off remained a mystery to America's nineteenth-century beekeepers.

Three decades later, an American beekeeper named R. C. Aikins wrote to the trade journal *Gleanings in Bee Culture* about a sudden, massive, and inexplicable die-off of the bees in his care.* Aikins, who witnessed this collapse in Fort Collins, Colorado, was profoundly disturbed and wrote, "It remains yet a complete mystery. Should it strike a whole State as it struck Denver last year, the consequences would be almost annihilation of the bee-business." Once again, scientists and officials remained uncertain about the factors that precipitated this disruption in hive health.

What is unique about the twenty-first-century threats to bee colonies is their breadth. While no single factor alone is sufficient to trigger colony collapse, several "sublethal" stressors can interact to destroy a colony. One of the prime suspects is the aptly named *Varroa destructor,* a parasitic mite that infests a honeybee colony's brood, composed of its eggs, larvae, and pupae. These tiny button-shaped mites enter the hive and suck the hemolymph (insect blood) of adult bees and their larvae. They also transmit to the bees viruses and fungi, some of which cause debilitating genetic defects like useless wings. Parasites and pathogens may work in concert to cause CCD. A controlled study performed in 2015 demonstrated that parasites, when combined with bee viruses, contributed to 70 percent of hive losses.

Recent research has shown that pesticides are also a key player in CCD. One class of insecticide of particular concern to scientists is the neonicotinoids, synthetic analogues of nicotine, a natural compound that plants such as tobacco synthesize to defend themselves. Prior to the Second World War, farmers in the United States had applied copious quantities of natural nicotine to their crops as an effective insec-

---

* In addition to stories on beekeeping, the illustrated semimonthly journal *Gleanings in Bee Culture* covered the eclectic array of editor Amos Ives Root's interests. In January 1905, it published the first accurate eyewitness account describing the flights of Wilbur and Orville Wright's airplane at Huffman Prairie, northeast of Dayton, Ohio.

ticide against the beetles, aphids, and insect larvae that plagued their crops. However, during the opening decades of the Synthetic Age, cheaper industrially produced neonicotinoids replaced their natural predecessor. Recent studies have shown that neonicotinoids have a range of unintended impacts on pollinators, which consume, spread, and inoculate their young with neonicotinoid-contaminated pollen and nectar. In addition to poisoning honeybees and killing them outright, pesticides like neonicotinoids make these insects more susceptible to parasites and pathogens.

A second-order problem with using synthetic insecticides is that neonicotinoids remain in the plant longer than natural insecticides do and can persist for extended periods in the soil because of their unusually slow rate of decay (some have half-lives as long as 1,155 days). The British environmentalist George Monbiot has famously dubbed neonicotinoids "the new DDT." This time around, the potency of these pesticides is orders of magnitude stronger than the chemicals Rachel Carson targeted in *Silent Spring*. As Monbiot has pointed out, by volume, "these poisons are 10,000 times as powerful as DDT." With the twenty-first century well under way, Carson's prescient warning from 1962 now seems more relevant than ever:

> These insects, so essential to our agriculture and indeed to our landscape as we know it, deserve something better from us than the senseless destruction of their habitat. Honeybees and wild bees depend heavily on such "weeds" as goldenrod, mustard, and dandelions for pollen that serves as the food for their young. . . . By the precise and delicate timing that is nature's own, the emergence of one species of wild bees takes place on the very day of the opening of the willow blossoms. There is no dearth of men who understand these things, but these are not the men who order the wholesale drenching of the landscape with chemicals.

Carson's prophetic words offered a foretaste of a contemporary news cycle that has begun acknowledging such threats to domesticated and wild bees.

In large part, this increased awareness is due to the emergence of a new breed of environmental advocacy organization in the last three decades of the twentieth century. Among the most robust examples of this shift is the Xerces Society for Invertebrate Conservation, based in Portland, Oregon. This international nonprofit, which promotes the conservation of invertebrates and their habitats, takes its name from the Xerces blue butterfly (*Glaucopsyche xerces*), the first North American butterfly known to have gone extinct as a result of human activities. The organization got its start in 1971 when a young lepidopterist and nature writer named Robert Michael Pyle returned from a Fulbright Fellowship in England, where he had studied butterfly conservation. Motivated by many of the warning signs that had spurred Rachel Carson to action, Pyle was determined to do something about the declining health of North America's insect population. Over the last half century, the society has focused its efforts on protecting the habitats of endangered invertebrates, advocating for dedicated "pollination corridors" to assist vital insect species in maintaining their migratory routes, training farmers and land managers in practical strategies for environmental conservation, and raising awareness about invertebrates (many of them pollinator insects) that inhabit the world's forests, prairies, deserts, and oceans.

As Pyle and the other members of the Xerces Society are quick to acknowledge, they are channeling deeper currents of environmental thought that flow outward from the nineteenth century or, in some cases, from wellsprings of Native American traditional ecological knowledge. One of these "streams of consciousness" is the writing and activism of Sierra Club founder John Muir. In the immediate aftermath of the Civil War, the twenty-nine-year-old Scottish-American naturalist set out alone on a thousand-mile trek through the rural South, wandering across Kentucky, Tennessee, North Carolina, South Carolina, Georgia, and Florida to the Gulf of Mexico. Along the way, Muir sketched scenes of the places he visited and kept vivid journal entries of his encounters with freed slaves, destitute Confederate soldiers, poor farmers, and Native American residents of the rapidly changing region. His descriptions of these interactions are intriguing in their own right. However, it is his exquisite accounts of interactions with nonhuman nature that animate his prose. After a night spent sleeping on the outskirts of Charleston, South Carolina, among the crumbling tombstones and moss-draped oaks of Bonaventure Cemetery, Muir awoke to find that "Large flocks of butterflies, all kinds of happy insects, seem to be in a perfect fever of joy and sportive gladness. The whole place seems like a center of life. The dead do not reign here alone."

As mid-nineteenth-century contemporaries, John Muir and Charles Darwin witnessed a remarkable confluence of scientific and social revolutions that took the world by storm.

8

A SIX-LEGGED MENU

In 2006, one of America's premier amusement parks offered a challenge to visitors. Declaring that "a cockroach is your ticket to the
front of the line," Six Flags Great America of Gurnee, Illinois, promised
celebrity treatment to anyone who volunteered to consume a live Madagascar hissing cockroach. The pre-contest press release channeled the

Madagascar hissing cockroaches (*Gromphadorhina portentosa*) are tawny,
ovular insects with shiny exoskeletons that resemble polished mahogany.
They are wingless, sport large antennae, and mature to a length of
two or three inches. They do not bite. Instead, when they are mating or feel
threatened, they forcibly expel air through respiratory holes called spiracles,
producing their trademark hissing sound. They are content dining on
a smorgasbord of carrots, fresh greens, and dog food and are common
house pets in Japan and parts of Southeast Asia.

melodrama of a late-night TV infomercial: "Daring guests willing to eat their way to the front of the line will have to be quick! Cockroaches can run at a speed of three miles per hour. The nutritional benefit? Cockroaches are extremely low in fat and high in protein." Despite its lighthearted tone, this remark captured the culinary Zeitgeist of the twenty-first century.

In recent decades, entomophagy—the eating of insects—has attained marquee status among certain food aficionados, scientists, and policy makers. The term "entomophagy" is relatively young. It first appeared in 1871 in the *Sixth Annual Report on the Noxious, Beneficial and Other Insects of the State of Missouri*. The report's author, state entomologist Charles Valentine Riley, was an iconoclast who developed a compendium of insect recipes to give beleaguered farmers a means of coping with the swarms of Rocky Mountain locusts (*Melanoplus spretus*) that invaded the Great Plains in the 1870s. Riley's buzzword remained obscure until the beginning of the twenty-first century, when environmentalists and gastronomes rediscovered its utility. Hailed for its potential to solve future food security problems and for its bold foray into unexplored flavors and textures, entomophagy is today experiencing a renaissance in the West.

Nutritionists and policy experts tout the health benefits and environmental sustainability of an insectivore diet. Insects convert what they eat into body mass much more efficiently than their larger animal counterparts, and they produce far fewer greenhouse gases than pigs or cattle. These ecological benefits are matched by the high nutrient content and low fat of edible insects. Comestible critters like mealworms, crickets, locusts, grubs, and ants are storehouses of healthy fats, protein, fiber, and minerals. To take just one example: grasshoppers pack the same amount of protein, two times the iron, and four times the amount of calcium as lean beef, while only containing a third of the fat.

Comparing land-use needs and water requirements for cows and crickets is revealing. The production of a pound of beef in the United States requires one thousand gallons of water and two acres of grazing land. In contrast, to raise a pound of crickets, you need only a gallon of water and two cubic feet of space. Raising livestock and growing the feed for cattle take up 70 percent of the world's agricultural land.

A 2010 report from the United Nations Food and Agriculture Organization announced, "Scientific analysis confirms . . . the exceptional nutritional benefits of many forest insects, and studies point to the potential to produce insects for food with far fewer negative environmental impacts than for many mainstream foods consumed today."

The popular press has favored more flamboyant enticements. A British writer for *The Independent* waxed poetic about the sensuous possibilities of six-legged gastronomy: "Ants that burst with honey as you bite into them. Giant hornet pupae that melt like cream on your tongue. Beetle larvae that leave a smoky taste in your mouth. And those are just the ones that can be eaten raw." Celebrity endorsements have lent a helping hand. As the actress Angelina Jolie told a BBC interviewer, "You start with crickets and a beer, and then you kind of move up to tarantulas." In similar fashion, international pop icon Justin Timberlake served a crispy buffet of ants and grasshoppers, "glazed in chocolate, garlic, and rosé oil," at the New York release party for his 2018 album, *Man of the Woods*. Timberlake's trays of insect hors d'oeuvres were styled by the Danish celebrity chef René Redzepi, whose Copenhagen restaurant Noma has won two Michelin stars for its audacious interpretations of Nordic cuisine.

Such high-profile promotional efforts have extended into the literary realm. Books with embellished titles like *Edible: An Adventure into the World of Eating Insects and the Last Great Hope of Saving the Planet* and *Creepy Crawly Cuisine: The Gourmet Guide to Edible Insects* typify a lively genre devoted to transforming flights of fancy—or night terrors—into earthly delicacies. The vibrant photos and graphic firsthand accounts that fill the pages of *Man Eating Bugs: The Art and Science of Eating Insects* are among the most colorful illustrations of this diversity. The book's authors, photojournalist Peter Menzel and writer Faith D'Aluisio, spent the better part of eight years traveling to more than a dozen countries in their quest to sample six-legged fare. Ugandan curried capricorn beetle larvae, Australian witchetty grub dip, and Chinese ant wine were among the locally popular insect dishes they tasted.

The movement toward global culinary tourism has exposed many Americans and Europeans to a wider world of food cultures with deep-seated traditions of entomophagy. Before his death in 2018, celebrity chef and travel documentarian Anthony Bourdain was responsible for sharing with a generation of television viewers the pleasures of munching on edible insects in all of their raw and cooked incarnations.

Regardless of their tactics, most European and American promoters of an insectivore lifestyle acknowledge the daunting taboos they face. Within Western cultures, aversion to entomophagy is hardly new. A nineteenth-century manifesto titled *Why Not Eat Insects?* shares many similarities with its twenty-first-century counterparts. Its author, the British entomologist Vincent M. Holt, wrote in 1885, "In entering upon this work I am fully conscious of the difficulty of battling against a long-existing and deep-rooted public prejudice. I only ask of my readers a fair hearing, an impartial consideration of my arguments, and an unbiased judgment. If these be granted, I feel sure that many will be persuaded to make practical proof of the expediency of using insects as food."* Holt's exhortations failed to alter the diets of his Victorian contemporaries. Even so, entomophagy was not unknown to Holt's peers.

Nineteenth-century recipe books frequently assumed that homemakers were familiar with a basic repertoire of insect ingredients. In 1840, the celebrated American cookbook author Eliza Leslie (popularly known as "Miss Leslie") recommended that her readers crush a pinch of dried cochineal bugs with a mortar and pestle to obtain "a good red colour" when pickling cabbage and preserving quinces and apples, adding that cake frosting could be tinted "pink by the addition of a little cochineal." Six years later, the American educator Catharine Esther Beecher advocated powdered cochineal as a full-bodied coloring for candies and desserts.

In Europe, nineteenth-century French chefs touted the virtues of *larves de hanneton sautées*, or dried cockchafer grubs (larvae of beetles in the genus *Melolontha*), known as May bugs or doodlebugs in English. One French recipe directed cooks to "place the live grubs in vinegar for several hours. Then dip in an egg, milk, and flour batter

---

* René Antoine Ferchault de Réaumur's six-volume *Memoires pour servir à l'histoire des insectes* (Notes to use in the History of Insects), 1734–42, and Foucher d'Obsonville's *Philosophic Essays on the Manners of Various Foreign Animals*, 1784, were two earlier attempts by Europeans to promote entomophagy among their compatriots.

and fry in butter. Or fry the live grubs in butter oil to which chopped parsley and garlic have been added." These sweet and savory insect concoctions notwithstanding, most Western societies have squeamishly resisted eating bugs.

It is a truism to suggest that a culture's food traditions are among its most celebrated features. The dishes, dining customs, and cooking techniques that distinguish our ethnic backgrounds and define our communities can seem timeless and immutable. Recognizing the primal role of these epicurean sensibilities, preeminent twentieth-century food writer M.F.K. Fisher remarked, "First we eat, then we do everything else." Such fundamental assumptions make it difficult, if not nearly impossible, to imagine how such tastes and traditions might ever shift. But rapid transformations have occurred.

Europe's change of heart about the potato (*Solanum tuberosum*) in the 1700s and Japan's swift embrace of a carnivorous diet at the end of the nineteenth century vividly illustrate how food preferences can undergo radical alterations. Although the potato has a truly ancient history in the Americas, its contributions to European cuisine are relatively recent. Around 3400 B.C.E., farmers on the steep slopes of the Andes domesticated wild tubers. Thousands of years before the arrival of Spanish conquistadors in the 1530s, these planters carefully selected from an abundant array of starchy root plants, cultivating as many as a thousand varieties of potatoes, whose flesh ranged in color from Corinthian purple to sunset orange. When Spaniards returned from the Americas with bags of swollen, knobby roots, the potatoes were not initially given a welcome reception.

These misgivings had merit. Potatoes are part of the Solanaceae family of flowering plants, better known as the nightshades. Many members of this botanical group—which includes tomatoes, eggplants, peppers, and tobacco—produce alkaloids, organic compounds that can be toxic to predators. Eating the green stems, berries, or leaves of nightshades can cause indigestion or more severe allergic reactions in humans. Because Europeans hastily borrowed these plants from Amerindian cultures that had developed long-standing customs of cultivation and consumption, misunderstandings abounded. The potato's reputation was also hindered by its unconventional propagation. It is not a seed plant like its starchy cousins, wheat and rice. Rather, it germinates underground from tuberous chunks, which sprout from "eyes," or small growing buds, on their surfaces. In 1748, the French government banned the cultivation of the gnarled spuds, which were rumored to cause fever, nausea, leprosy, and an uncontrollable swelling of the glands.

The undoing of this anti-potato prejudice was the work of a clever French physician. Along the marble ledges of an ornate burial monument in the Père Lachaise cemetery of Paris, visitors have placed rows of golf ball–sized brown potatoes. These vegetal tributes pay homage to Antoine-Augustin Parmentier, champion of the humble spud. Born in 1737, Parmentier eventually became a member of the French army's pharmaceutical service. Having been taken prisoner five times by the Prussian army during the Seven Years' War (1756–63), Parmentier had eaten countless prison meals of the Andean cultivar that was fast becoming the centerpiece of central European diets. He concluded that hardy, energy-dense potatoes could also help the French peasantry endure the famines that periodically ravaged his homeland.

Parmentier was an unrivaled master of publicity. He offered sumptuous bouquets of purple potato flowers to King Louis XVI and Queen Marie Antoinette, and he served potato-based dishes—soups, liquors, and everything in between—at banquets for notable foreign guests, including Benjamin Franklin and Thomas Jefferson. In 1789, on the eve of the French Revolution, Parmentier published his *Treatise on the Culture and Use of the Potato, Sweet Potato, and Jerusalem Artichoke.* "Printed by order of the King," the pamphlet affirmed, touting the royal endorsement of potato eating. Republicans also accepted its groundbreaking premises, making the potato one of the few constituents of the ancien régime to survive the guillotine. Today's cookbooks and culinary websites abound with recipes for "Parmentier potatoes," cubed spuds baked in a dish with sautéed herbs and butter.

Japanese history offers a similar tale of dietary upheaval. The food cultures of medieval Japan prioritized vegetarian fare, supplemented by fish and fowl. Over many centuries, the Buddhist precept against the killing of sentient beings—creatures that actively seek pleasure and avoid pain—decisively shaped the country's food preferences. In 675 C.E., the devout Emperor Tenmu enforced this dictum by banning the consumption of cattle, horses, dogs, chickens, and monkeys. His policy endured, in various guises, for nearly twelve hundred years.

Meat prohibitions began to falter toward the end of Japan's Edo period (1603–1868). Encounters with mammal-eating Westerners made a lasting impression on Japanese policy makers. In 1869, during this new era of contact with the West, the reformer Kakuta Tôru pointed out that the British "frequently eat beef" and "rarely get fatigued in the battlefield." The association between carnivorous diets and the nourishment of soldiers' stamina became axiomatic for Japanese military commanders.

At the same time, Japanese administrators relaxed the enforcement of nationwide meat-eating prohibitions. The changing tastes of imperial officials played a role in this transition. Tokugawa Yoshinobu (1837–1913), the last *shōgun*—hereditary commander in chief—of feudal Japan expressed such a steadfast fondness for pork that he acquired the nickname Ton'ichi-sama, or "pork-loving *shōgun* of the Hitotsubashi clan." Specialty restaurants began offering hunted game to wealthier patrons for "medicinal eating." In major cities, diners could walk into "beast markets" and order dishes such as wild boar, sold under the mysterious euphemism "mountain whale" (*yamakujira*). Likewise, red meat began appearing on menus as "winter peony" (*fuyu botan*). These semantic disguises offered pathways of permissibility to formerly taboo foods. On January 24, 1872, a formal announcement confirmed that Japan's Meiji emperor consumed beef and mutton on a regular basis. This declaration ended centuries of meat embargoes.

It remains to be seen whether insect eating will experience revolutionary transformations in Western culture similar to the acceptance of potato consumption in Europe and the emergence of a carnivorous diet in Japan. However, there has never been any shortage of footholds for insectivores in many cultures the world over. Some of these underground food traditions have thrived in unexpected places.

On the Mediterranean's second largest island (after Sicily), Sardinia's locals indulge in one of the world's illicit gastronomical pleasures. *Casu marzu*, literally "rotten cheese," is the piquant outcome of leaving a wheel of sheep's-milk pecorino to ripen in a dark hut for several months. The runny aftermath hosts a squirming larval colony of cheese flies (*Piophila casei*). In a distinction lost on all but the most die-hard aficionados, the presence of live maggots is among the telltale signs that the pecorino is properly aged, not spoiled. These translucent white worms, about the size and shape of grains of rice, are renowned for their ability to leap several inches. Locals recommend that first-timers spread their gooey, squirming cheese on a slab of moistened flatbread and close their eyes when taking bites of the ammoniac delicacy. This time-honored ritual is often accompanied by a strong red wine such as *cannonau* to keep matters flowing smoothly across the palate, through the gut, and out of the bowels. Despite a European Union ban on *casu marzu* for its violation of established hygienic standards, consumption of the larval cheese thrives throughout Sardinia, where locals celebrate this ritual as a pungent reminder of their cultural heritage.

Across the Mediterranean, a few thousand miles to the southeast of Sardinia, is the homeland of the Bible's most celebrated insect eater. The Gospel of Mark introduced John the Baptist, the baptizer of Jesus, as a frugal entomophagist. Clad only in a camel's-hair robe and a leather belt, John survived the inhospitable Judean wilderness on a diet of honey and locusts. His ascetic insectivore regimen followed the dietary patterns of his distant ancestors. As evolutionary anthropologists are quick to point out, more attention should be given to the role of eating insects in the development of humans. Evidence is mounting that a smorgasbord of beetles, ants, bees, termites, butterflies, moths, locusts, and crickets was among the most sought-after treats of hominid foragers.

Idealized images of hunter-gatherers have animated a contemporary food craze known as the caveman diet, also called the paleolithic diet. Emboldened by the American nutritionist Loren Cordain's bestselling 2002 book, *The Paleo Diet,* adherents to this Stone Age strategy eat foods that were available to our human forebears before the widespread domestication of grains some ten thousand years ago. Proponents of this diet contend that humans have not experienced the vast evolutionary intervals necessary to evolve a gut capable of effectively digesting the grain-based fare that accompanied the agricultural revolutions and more sedentary lifestyles of modern civilization. Therefore, bodies outpaced by hasty innovations in food production create a devastating biological mismatch that contributes to obesity, diabetes, heart disease, and many of the other maladies of affluence that plague the developed world. The paleo diet's Pleistocene menu prioritizes the consumption of lean meats and fish, berries, vegetables, nuts, and seeds. It stresses the avoidance of grains, dairy products, refined sugar, potatoes, and processed foods.

Such claims have some validity. Diets that eliminate artificial additives, abolish excess sugar, and shun salt and cholesterol-rich ingredients are likely to produce better health outcomes for their adherents. Most humans benefit from consuming lean proteins and high-fiber fruits and vegetables, which serve as cornucopias of vitamins and minerals. In addition, *Homo computus,* hunched over a glowing screen, scarfing down greasy potato chips, microwaved burritos, and syrupy soft drinks makes for a compelling symbol of contemporary social decay.

Even so, there are reasons to be skeptical of some of the less credible assertions behind the paleo fad. The American evolutionary biologist Marlene Zuk has called such nostalgia for prehistoric fare a "paleofantasy" predicated on fundamental misunderstandings of human dietary evolution. Scientists have come to understand that our distant ancestors were far less like today's barbecue pit masters, with their slabs of ribs, and far more like Popeye the Sailor with his cans of spinach. For the most part, our predecessors bulked up on greens, not flesh. Mounting evidence suggests that their diets revolved around substantial helpings of chlorophyll-rich wild vegetables.

While it is tempting to imagine roving bands of fur-clad cavemen pelting woolly mammoths with a hail of spears and rocks, meat from large mammals was a comparatively rare indulgence. Indeed, most of the supplemental animal protein that hominids consumed came from insect bodies. According to researchers Stanley Boyd Eaton and Dorothy A. Nelson, "From the time mammals first appeared until 50 million years ago—a total of 150 million years, three quarters of the entire time mammals have existed—our ancestors were primarily insectivorous." Geographical variety certainly shaped the dietary options available to our prehistoric relatives, but bugs seem to have gotten top billing on the menu whenever and wherever possible.

Restoring insects to their central position in the human diet is the core mission of several dozen start-up businesses that have emerged in the past decade. All Things Bugs, Bitty Foods, Bud's Cricket Power,

Chapul, CRIK Nutrition, Exo Protein, Lithic Foods, and Six Foods are just some of the catchy names for the innovative enterprises now producing a dizzying array of products fortified with high-protein cricket meal. Made from frozen crickets that have been dry-roasted and milled into a powder, the "flour" has a mild, nutty taste. It is ground so finely that no antennae or legs protrude from the pulverized product, thus disguising any evidence of its origins. The lavishly illustrated websites of cricket flour companies suggest a miscellany of cooking applications and recipes ranging from six-legged smoothies to cocoa chirpy brownies, cricket protein spicy omelets, and ricotta cricket pancakes. A 2019 report by the financial services firm Barclays predicted that edible insects will become an $8 billion business by 2030.

In addition, a new breed of advocacy organization has emerged to broadcast the virtues of entomophagy. Since its creation in 2013, Little Herds, based in Austin, Texas, has championed edible insects as a source of stability in an increasingly precarious world. Targeting open-minded skeptics, this educational nonprofit informs the public about the environmental and dietary benefits of high-protein "mini-livestock" that can be sustainably grown in urban areas. Little Herds runs pitch competitions for bug-based marketing ideas, helps farmers use fly larvae to turn food scraps into chicken feed, and produces entomophagy teaching kits for elementary school classrooms.

While organizations in the United States and Europe work to overcome psychological taboos that limit insect consumption, cultures throughout the rest of the world continue their venerable traditions of entomophagy. At least 113 countries on six continents embrace long-standing practices of insect eating.

Such customs have thrived for millennia in many Asian countries, including Korea, Thailand, and Japan. Among the many distinctive smells that pervade a traditional Korean marketplace is the tangy, fish-like aroma of *beondegi*. With a name that translates to "pupa," this popular street food is prepared from silkworm chrysalids simmered in broth or fried in oil and spices. Vendors serve heaping paper cups of the steaming pupae, which customers eat with toothpick skewers. The chitinous bodies of the immature silkworms exude a briny-earthy flavor, akin to a curious cross between a shrimp and a peanut.

Elsewhere in Asia, the giant water bug (*Lethocerus indicus*) retains its long-term status as a wildly popular edible insect. In northern Thailand, this palm-sized creature, which looks like a bronze-colored cockroach with large foreleg pincers, is a common component of street fare and restaurant cuisine alike. Thai cooks often deep-fry the larger females to be eaten whole, while they crush the males with a mortar and pestle and mix them with chilis, onions, and garlic to make the piquant relish, *nam prik mang da*. When the bodies of the male bugs are ground, their pheromones impart the musky aroma of ripe flowers. Much like China's Sichuan peppercorn (*Zanthoxylum bungeanum*), *nam prik mang da* causes a slight numbing sensation on the consumer's tongue.

Japan also sustains long-standing traditions of entomophagy. Along the upper reaches of the Tenryū River, a major watercourse that snakes through the heart of Japan's Honshū Island, fishermen deftly manipulate bamboo-framed scoop nets (*yotsude ami*) to snag squirming reddish-brown bundles of aquatic fly larvae that cluster among the rocks on the river bottom. These *zazamushi*—literally "insects under the slow current"—are considered a rare delicacy. Chefs boil the larvae and then sauté them with soy sauce and sweet rice wine or sugar.

Some traditional African cuisines likewise feature six-legged dishes. Eight thousand miles to the southwest of Japan, the larvae of another moth species star in a much-sought-after treat. Thick as a cigar and as long as a human hand, the caterpillars of the mopane emperor moth (*Gonimbrasia belina*) sustain a multimillion-dollar industry in Botswana, Namibia, South Africa, Zambia, and Zimbabwe. The mopane caterpillar, a species native to the semi-arid regions of southern Africa, is recognizable from its characteristic pattern of alternating whitish-green and yellow stripes, spiked with short black or reddish spines covered in fine white hairs. A vast cottage industry is propelled by the labors of the women and children who gather the mopane caterpillars by hand, squeeze out their entrails, and boil the bodies in salt water, before spreading them out to dry in the sun. Depending on how they are prepared, the insects can exude the tang of black tea, suggest the consistency of jerky, or taste like a well-done steak. Dehydrated mopane caterpillars can last up to a year, a virtue in hot regions where refrigeration is often in short supply.

Much like the traditional ecological knowledge required for cul-

The caterpillars of the mopane emperor moth (*Gonimbrasia belina*) are considered a nutritious snack and sustain a multimillion-dollar industry throughout much of southern Africa.

tivating shellac insects, silkworms, and cochineal bugs, an intimate understanding of the plants that host mopane caterpillars is required for raising these southern African moths. Mopane caterpillars feed exclusively on the butterfly-shaped leaves of mopane trees (*Colophospermum mopane*), and some villagers supplement their incomes by renting their groves to caterpillar collectors during the austral summer mopane harvest.*

The mopane caterpillar has wiggled its way into literature and politics. Mma Precious Ramotswe, the fictional private eye from the Anglo-Zimbabwean writer Alexander McCall Smith's popular novels about Botswana's No. 1 Ladies' Detective Agency, is a compulsive consumer of dried *mopanis*.† In real life, Malawi's iron-fisted president

---

* Mopane trees are also the food plant for the reddish-brown wild silk moth, *Gonometa rufobrunnea,* which produces cocoons that can be harvested and spun into threads.

† In Setswana, one of Botswana's two official languages, "Mma" is a term of respect—much like "Mrs." or "Madam"—used before a woman's name. The male version (like "Sir" or "Mr.") is "Rra."

Hastings Kamuzu Banda (1898–1997) famously gorged on mopane caterpillars. He carried them around in his pockets as treats for the children he encountered on his frequent visits to the countryside during his thirty years in power.

Half a world away, insects also play important roles in many New World cuisines. Eager insect consumers throng the bustling outdoor marketplaces of Mexico. Cinnabar-red *chapulines,* toasted grasshoppers that have been dusted with chili powder, sautéed in garlic oil, and showered with fresh lime juice, are one of southern Mexico's most distinctive and coveted delicacies. Residents of this region have feasted on *chapulines* since long before the Spanish arrived in the early sixteenth century, and they still use traditional cooking techniques, roasting these six-legged treats on a smooth iron griddle known as a *comal.* Anna Garcia, a farmer from Oaxaca's rural Central Valleys, describes the harvest: "If you try to catch the bugs in the middle of the day, you look like an idiot jumping around, and you can't catch them anyway. It is too hot. You need to go out early with a net to get them."

Undeterred by national borders, *chapulines* have been migrating northward. Fried grasshoppers are not a typical item on the traditional U.S. ballpark menu of hot dogs, peanuts, beer, and popcorn. However, when Manny Arce, the executive chef of Seattle's Poquitos restaurant, approached the management at Safeco Field—the home of the Seattle Mariners baseball team—with the idea of adding these Oaxacan treats to the game-time offerings at his food stand, stadium administrators agreed to the proposal. Ever since *chapulines* debuted on the menu in 2017, the spicy insect appetizers have received a surprisingly warm welcome from the park's visitors. In the words of one fan, "I think one thing most people are afraid of when they imagine eating insects is that they'll be squishy or gooey inside. But the chapulines were light and crunchy, and mostly tasted like lime with a hint of chili." Arce sold more than eighteen thousand orders of spicy, toasted grasshoppers during the first two weeks of the 2017 baseball season.

Other Mexican insect delicacies have yet to enter mainstream cuisine north of the Rio Grande. *Escamoles,* known as "Mexican caviar," are the larvae of velvety tree ants (*Liometopum apiculatum*) native to the states of Guanajuato, Puebla, Hidalgo, and Tlaxcala. They make a delicate accompaniment to hearty entrées like *barbacoa* (meat steamed in an underground oven) and *mixiote* (oven-baked, marinated meat wrapped in banana leaves). *Escamoles* resemble corn kernels and have a nutty, buttery taste. Since they can be collected only during March and April and arduous labor is required to extract them from their underground tunnels beneath prickly maguey plants in the high plains of central Mexico, these prized commodities retail in Mexican specialty markets for as much as fifty dollars a pound.

Such traditions are hardly new. In fact, insect eating was a mainstay of the ancient foodways of the Americas. Entopreneurs, as today's bug-food innovators call themselves, have capitalized on such deep-seated historical legacies. As one Bay Area start-up advertises: "Don Bugito, the Prehispanic Snackeria, is a San Francisco based company focused on planet-friendly protein snacks, featuring delicious edible insects in savory and sweet flavors. . . . Don Bugito focuses on treats inspired by Pre-Columbian Mexican cuisine and is always working toward re-inventing ancestral food." Such promotional strategies are not hyperbolic. The Florentine Codex, a sixteenth-century manuscript compiled by the Spanish Franciscan friar Bernardino de Sahagún, records many traditional Mesoamerican insect-eating practices. For example, Sahagún and his Native American Nahua co-authors noted that Aztecs prized the maguey caterpillar (*Aegiale hesperiaris*), or *meocuili,* which they packed in leaf pouches and then crisped on bakestones or roasted over glowing embers. Today, red maguey larvae are one of the types of *gusanos* (worms) found in bottles of Oaxaca's famed mezcal liquor.

Dozens of insect species also featured prominently in the highly developed culinary traditions of North America's Native American communities. The Onondaga, one of the original five nations of the Iroquois Confederacy, used ants to add a citric zest to various recipes. As one early-twentieth-century observer put it, the Onondaga consumed them "more as a luxury than as a staple." These creatures were not "famine foods," used to eke out survival in times of deprivation. Rather, they were chosen for their delectable flavors.

Elsewhere, Native American communities used edible insects to add dietary variety, complement protein supplied from wild game hunting, and offer a reliable source of nutrition. Nineteenth-century ethnographer Walter J. Hoffman described the massive grasshopper harvests organized by the Shoshone and other Indian communities in the American West:

> A fire is built covering an area of from 20 to 30 feet square, and as the material is consumed to coals and ashes, all the Indians start out and form an extensive circle, driving the grasshoppers with blankets or bunches of brush toward the centre, where they are scorched or disabled, [and then] they are collected, dried, and ground into meal. With the addition of a small quantity of water this is worked and kneaded into dough, formed into small cakes, and baked in the sand under a fire.

Such practices were the distant forerunners of today's cricket-flour start-ups.

In the mid-nineteenth century, Native American traditions of entomophagy frequently sustained whites during their westward expeditions. Edwin Bryant, a Kentucky newspaper editor who journeyed overland to California in 1846, wrote of exchanging sewing needles for food provided by Native Americans of the Great Basin. His account mixed condescension with a reluctant acceptance of the importance of these insects to his own party's survival. At one point, several Shoshone women offered the travelers some loaves, "which, upon examination we ascertained to be service-berries, crushed to a jam and mixed with pulverized grasshoppers. This composition being dried in the sun until it becomes hard, is what may be called the 'fruitcake' of these poor children of the desert." As Bryant grudgingly admitted, "The prejudice against the grasshopper 'fruit-cake' was strong at first, but it soon wore off, and none of the delicacy was thrown away or lost."

Native American insect-eating traditions have survived deep-seated cultural prejudices and persist to this day. Semiannually, members of the Big Pine Paiute Tribe of California's Owens Valley dig trenches around conifers to harvest Pandora moth (*Coloradia pandora*) caterpillars. After they smoke the larvae out of the forest canopy, men and women gather the fallen caterpillars, then roast and boil them to be served unaccompanied or as an ingredient in a flavorful meat-and-vegetable stew.

Given how commonplace entomophagy has been throughout most

of the world, what explains the widespread aversion to such practices in most of Western society? According to the renowned British anthropologist Brian Morris, "There is no evidence at all that humans have an instinctive dislike of insects as food. On the contrary, throughout human history insects of many kinds have been eaten by human communities, not simply as a 'famine' food but as an intrinsic part of their diet." Some of the bias against eating insects may have arisen from the fact that the European colonists of North and South America frequently came from colder, northern latitudes where edible insects were not abundant and had never been traditional culinary mainstays.

The racial prejudices that served as handmaidens of colonialism shaped associations between food preferences and supposed savagery. After describing the frequent consumption of locusts by the Kumeyaay people of Baja California, the Mexican Jesuit priest Francisco Javier Clavijero wrote that they were being dissuaded from such habits by the "good advice of the missionaries and the experience acquired in 1772, in which a great epidemic attacked the Indians because they ate so many locusts." In all likelihood, this eighteenth-century epidemic was a smallpox outbreak that had nothing to do with Kumeyaay dietary preferences.

Such biases survived into the twentieth century. In his 1946 book, *Insects, Food, and Ecology,* the Harvard entomologist Charles Thomas Brues unabashedly racialized the boundary between those who eat insects and those who do not: "Even man himself does not completely exclude insects from his dietary, although the consumption of insects by members of our western civilizations is at the present time entirely unintentional on their part. Such is not true of other peoples, and if space permits we may return to this as it serves royally to bolster up the feeling of race superiority, Nordic or otherwise." Ironically, the "other peoples" who depend upon insects as a fundamental protein source may be evading many health risks.

A noxious brew of pathogens festers in today's confined animal production facilities, awaiting opportunities to jump from animal populations to human hosts. Factory-style chicken farming has fostered an upsurge of antibiotic-resistant bacteria, while industrial-scale hog farms are colossal petri dishes of pandemic-causing influenza viruses. Beyond the animal cruelty of such production practices, against which

activists like English chef Hugh Fearnley-Whittingstall have successfully campaigned, they pose serious health risks. In 2011, the Dutch
entomologists Marcel Dicke and Arnold van Huis pointed out in
The Wall Street Journal, "Raising insects for food would avoid many
of the problems associated with livestock. For instance, swine and
humans are similar enough that they can share many diseases. Such
co-infection can yield new disease strains that are lethal to humans, as
happened during a swine fever outbreak in the Netherlands in the late
1990s. Because insects are so different from us, such risks are accord-
ingly lower." Starting in 1971 with Frances Moore Lappé's best-selling
Diet for a Small Planet, food writers have often described eating as a
political act. What has now become clear is that it is also an epidemio-
logical wager.

In many ways, the debate over the virtues of eating insects is moot.
We are all entomophagists, whether we know it or not. The most
devout vegan who shuns all animal-derived products may be alarmed
to learn that the United States Food and Drug Administration (FDA)
has deemed it acceptable for peanut butter to contain an "average of
30 or more insect fragments per 100 grams." FDA regulations permit twice that amount for chocolate, and government food inspectors
allow ground cinnamon to host up to 400 insect parts per 50 grams.
Even your daily cup of morning joe is a percolating mélange of mandibles and antennae. Shipments of unroasted coffee beans frequently
contain up to 10 percent insects. Like the cochineal in our candy and
cocktails or the shellac that coats our aspirin and shiny apples, insects
and their secretions flow in and out of our bodies on a daily basis.

Yet what role will *intentional* entomophagy play in the future of
food? Human diets in the twenty-first century will be deeply affected
by the economic constraints and biological limits of a hotter, drier
Earth as our climate continues to warm. Our planet is also becoming
a more crowded place. The United Nations predicts that the world's
population will reach 9.7 billion by 2050. Intensifying climate change,
coupled with the addition of two billion human beings, will pose a host
of new challenges to our existing global food system.

Among the few probable features of these uncertain times will be
the continued role of cereal grains like rice, wheat, corn, millet, and
sorghum as the starchy centerpieces of most diets. Half of the world's

current population, or more than 3.5 billion people, depend on rice for a fifth of their daily caloric intake. This is hardly a new trend. Rice has fed humanity for many millennia. It is impossible to overstate our shared dependency on the seeds of a tenacious grass from the genus *Oryza,* which humans domesticated independently in Asia, Africa, and South America.

Wheat will also be a staple of the world's meals, although not always in forms that would be recognizable to diners even a century ago. One of these novel incarnations is a four-by-four-by-one-inch block of curly, blond precooked noodles. Invented after World War II by an innovative Taiwanese-Japanese businessman named Momofuku Ando, these flash-fried bundles of flour, starch, water, and salt offered a culturally familiar way for Japanese to eat wheat rations supplied by U.S. occupying forces. As the anthropologists Deborah Gewertz, Fred Errington, and Tatsuro Fujikura point out in *The Noodle Narratives,* consumers around the world slurped an astounding 95 billion bowls of instant ramen in 2010.

Governments the world over are betting that insects will be an increasingly popular topping for all of this rice and wheat. Ultramodern design trends have played major roles in these gambles. Recently, Stockholm, Taipei, and London hosted large-scale displays of the Buzz-Building, a translucent tubular cricket-farming facility designed by the award-winning Swedish firm Belatchew Arkitekter. The luminescent glass-and-steel exoskeleton of this winding loop accommodates 111,406 square feet of farmable surfaces, an entomophagy-themed restaurant, an interactive corridor to give visitors windows onto the entire cricket-raising process, and a refuge for bees and other beleaguered migratory pollinators. BuzzBuilding is a leading exemplar of "agritecture," a design approach that focuses on urban food production. In a world where the majority of people live in cities, metropolitan farming has many advantages, ranging from more efficient land use to reductions in long-distance transportation of food.

Despite the justifiable hype around entomophagy, many questions remain unanswered. Åsa Berggren, a conservation biologist at the Swedish University of Agricultural Sciences in Uppsala, noted in 2019, "A lack of basic research on almost all aspects of production means the future environmental impact of the mass rearing of insects is largely unknown." Among the list of possible caveats are issues of scale. While small-sized insect farming operations are already making strides in the areas of land use and energy efficiency, such results do not always prevail when operations enlarge. As the authors of a 2015 study showed, the industrial expansion of cricket production leads to increases in the feed conversion ratio—the amount of food needed for an organism to reach harvestable size—thereby negating some of the sustainability "selling points" of entomophagy.

The mass production of insects for food may also generate other inefficiencies. Unlike meat, which is frequently sold raw (think of the rows of fat-marbled steaks, mountains of ruddy ground beef, and stacks of skinless chicken breasts that line refrigerated butcher cases), edible insects are generally raised in heated warehouses. Workers grind, dehydrate, and freeze-dry insect bodies to make them palatable to consumers. Each of these steps requires corresponding investments of energy, frequently derived from coal-fired power plants.

At a further remove, the portrayal of entomophagy as a one-off solution to world food shortages conceals the larger structural inequalities that caused those problems in the first place. There are no "free lunches" in the struggle to break the chronic cycles of nutritional deficiency, famine, and starvation. As the Nobel Prize–winning Indian economist Amartya Sen revealed in his detailed study of the 1943 Bengal famine, the deaths of as many as three million people were not caused by absolute shortages of food. In fact, the rice harvest that year was only marginally lower than it had been in 1942. Instead, as Sen demonstrated, undemocratic conditions were a prime culprit. Hunger had stemmed from "an entitlement failure," a situation where food supplies were inaccessible to the poor and unemployed. These populations lacked adequate social and economic power to assert their right to nourishment.* While the future of food will most likely involve the human consumption of insects, problems of unequal access to food will not be solved by a "silver bullet" with wings, antennae, and six legs.

---

* Irish economic historian Cormac Ó Gráda has supplied a contrasting view of the Bengal famine. He assigns less blame to market failures, arguing instead that India's British colonial rulers were unwilling to divert food from their wartime efforts against the Japanese.

Worlds apart from these daily struggles for sustenance, scientists are envisioning a pivotal role for entomophagy in interstellar exploration. A team of Japanese researchers has concluded that edible insects— with their resilient physiques, rapid reproductive rates, and nutrient-dense bodies—are among the most viable food sources to sustain a human expedition to Mars.

If the notion of interplanetary insect voyages sounds ludicrous, it is not. The first animals launched into outer space were neither monkeys nor dogs; they were fruit flies. The same species (*Drosophila melanogaster*) that has acted as our compass in navigating the mysterious topography of our own genetic landscape also led us into "the final frontier." On February 20, 1947, the United States military sent a capsule of fruit flies hurtling toward outer space aboard a captured Nazi V-2 rocket. Scientists launched the ballistic missile (little more than a huge pipe filled with ethanol and liquid oxygen) from White Sands Missile Range, in New Mexico. The V-2 reached an altitude of sixty-eight miles in 190 seconds. At that point, a pod broke away from the rocket, unfurled a parachute, and lowered its unwitting crew of six-legged astronauts down to the desert floor. The investigators were relieved to find that the hapless insects had survived the journey without noticeable harm.

Back on Earth, the mundane quest to secure the next meal is the top priority for most of humanity. In the coming decades, entomophagy is bound to play an increasing role in addressing this unending challenge. Yet there is an abundance of cautionary tales to temper the hype about an all-purpose protein concocted from freeze-dried crickets and powdered mealworms. Miracle substances like algae and soybeans have made similar cameos in past narratives about the ultimate solution to global hunger. In both cases, the results fell far short of the grandiose predictions that preceded them. Despite glowing prophecies in the 1950s and 1960s about the prospects for "chlorella cuisine" and soy-based diets, most of the planet's human population is not munching on algae burgers or dining on tofu turkey. In fact, the current reality is a disturbing inversion of such a scenario. As food writer Christine M. Du Bois has pointed out, "Some 70 percent of all soy protein in the world is used in the inefficient process of providing humans with protein in pork and poultry." Such wasteful food chains are gobbling up resources at an unsustainable rate.

Edible insects offer one of many possible routes toward a different future, but such a dietary shift would depend on changes to deeply ingrained habits and biases. However, when examined closely, the antipathy toward entomophagy that persists in the United States and Europe rests on fragile foundations. The Israeli entomologist Friedrich Shimon Bodenheimer once noted, "The aversion to insect food in Western civilization, though an established fact, is nevertheless not based on a hereditary instinct. It is established by custom and prejudice." Intergenerational transformations in behaviors and preconceptions are among the most reliable trends chronicled by historians. It is not far-fetched to imagine that the twenty-first century's consumers may respond differently than their nineteenth- and twentieth-century predecessors did to Vincent M. Holt's 1885 provocation: "Why not eat insects?"

EPILOGUE

LISTENING TO INSECTS

What does it mean to listen to insects? The renowned English diarist Samuel Pepys (1633–1703) recounted how his countryman Robert Hooke, a natural philosopher, was "able to tell how many strokes a fly makes with her wings (those flies that hum in their flying) by the note that answers in musique during their flying." Indeed, Hooke's ability to recognize the distinctive pitches emanating from various airborne flies has been verified by subsequent research. Like a sonic finger-print, the wingbeat frequency of almost every species of flying insect leaves a unique acoustic trace. This one-of-a-kind buzz, hum, or drone reveals the creature's identity. In the field of human health, wingbeat identification of insects has taken off as a way to detect the presence of disease-carrying mosquitoes. Using a sensor equipped with a laser beam, scientists can measure the wingbeat frequency of minuscule airborne fauna to determine which insects are present in any given environment, from a savannah in Tanzania to a swamp in Louisiana.

Undeniably, the whines and buzzes of bugs have been the harbingers of plagues and pestilence throughout human history. Nevertheless, in other times and places, their sonorous chirps, trills, and hums have been more pleasing to the ear. An ode to cicadas by the first-century B.C.E. Greek poet Meleager of Gadara artfully expresses this attitude:

O, shrill-voiced insect; that with dewdrops sweet,
Inebriate, dost in the desert woodlands sing;

Perched in the spray-top with indented feet,
Thy dusky body's echoing harp-like ring.
Come, dear cicada, chip to all the grove,
The Nymphs and Pan, a new responsive strain;
That I, in the noontide sleep, may steal from love,
Reclined beneath the dark overspreading plane.

The ancient Greeks were not alone in their high regard for insect songs. On autumn evenings during Japan's Edo period (1603–1868), couples would flock to Tokyo's Dōkan Hill, a popular spot for spreading picnic blankets, sipping sake, and listening to the bug concerts (*mushi-kiki*) of chirping crickets and sonorous cicadas.

Half a world away and several centuries onward, the Amazon River basin's nocturnal soundscape set the stage for Andrew Revkin's biography of the Brazilian rain-forest defender Chico Mendes: "The cicadas began their nightly drone, enfolding the town and the surrounding rain forest in a blanket of sound that resembled an orchestra of sitar players tuning their instruments." In such settings, the melodic rituals of insects offer comfort and stability in a turbulent world.

Insect sounds also resonate with human passions. A third of the way through "Touch Me," poet laureate Stanley Kunitz reminisced about one such moment:

Outdoors all afternoon
under a gunmetal sky
staking my garden down,
I kneeled to the crickets trilling
underfoot as if about
to burst from their crusty shells;
and like a child again
marveled to hear so clear
and brave a music pour
from such a small machine.
What makes the engine go?
Desire, desire, desire.
The longing for the dance
stirs in the buried life.

One season only,
        and it's done.

In his inimitable style, Kunitz braided his description of this fleeting interspecies encounter from wild, enchanting, and melodic lines.

Human responses to insect sounds are the products of particular cultural contexts. Still, there is a transcendent element that unites all of these different examples. Iconoclastic philosopher and avant-garde jazz artist David Rothenberg proposed that insect sounds were the forerunners of the fundamental elements in human music. In his 2013 book, *Bug Music: How Insects Gave Us Rhythm and Noise,* he contended that "the stridulation of crickets, the tymbaling of cicadas, the *tap-tap-tap*-ing of treehoppers . . . may be the very source of our interest in rhythm, the beat, the regular thrum." Rothenberg's suggestion is both tantalizing and persuasive. The buzzing vibrato that punctuates the Russian composer Nikolai Rimsky-Korsakov's "Flight of the Bumblebee" and the brisk hum of the fly-inspired melodic line in Edvard Grieg's "The Gadfly Said to the Fly" fittingly illustrate such assertions.*

Whether or not we are "in the groove" to beats with bug beginnings, it is clear that insects are sending us audible messages about our environment. Our six-legged cousins are extraordinary barometers of seemingly imperceptible changes in their surroundings. Counting a cricket's chirps for fourteen seconds and adding forty to this number is a surprisingly reliable way to approximate the current temperature in degrees Fahrenheit.

Insect messages about changes in the weather can also have more gravity. In fact, bugs often speak to us without ever making a sound. In Barbara Kingsolver's 2012 novel, *Flight Behavior,* twenty-eight-year-old Dellarobia Turnbow is enraptured by the sight of the fifteen million monarch butterflies that have come to roost in the woods near her small southern Appalachian town of Feathertown, Tennessee. Eventually, we learn that this miraculous occurrence is actually a symptom of an alarming phenomenon. The migrating colony of monarchs is wildly

---

* In 1960, Smithsonian Folkways Recordings released *Sounds of Insects,* a vinyl LP filled with insect songs and sounds and narrated by entomologist Albro T. Gaul.

off course, its centuries-old migration pattern altered by the chaotic weather of a warming planet.

The premises of Kingsolver's fictional scenario are corroborated by real-world science. In a 2019 study, ominously titled "Worldwide Decline of the Entomofauna," researchers conducted a wide-ranging survey of seventy-three case studies from around the globe and concluded that more than 40 percent of insect species face immediate danger of extinction. The authors determined that pesticide-intensive agriculture, habitat loss from urbanization, the rapid spread of pathogens and invasive species that destroy insect habitats, and the unpredictable weather patterns of Earth's changing climate were the most significant drivers of these declines. Henry David Thoreau considered "the hum of insects" to be "the evidence of nature's health or *sound state*." In today's rapidly changing world, his nineteenth-century insight seems more germane than ever.

A misguided commitment to bug eradication was a resolute policy goal during the first half of the twentieth century. "The Insects Are Winning: A Report on the Thousand-Year War," declared a featured article in the March 1925 edition of *Harper's Magazine*. The essay, written by William Atherton DuPuy, the executive assistant to the secretary of the Department of the Interior, argued: "The issue is vital; no less than the life or death of the human race. If man wins, he will remain the dominant species of this earth. If he loses, he will be wiped out by this, his most ambitious racial enemy." An overwhelming accumulation of data indicates that humans are winning this war, to our detriment and the impairment of fundamental environmental processes. Indeed, our victories to date have been Pyrrhic. They may turn out to be suicidal, for without insects, it will be extremely difficult to sustain human life on planet Earth.

It is impossible to know precisely what the future holds for the many long-standing antagonisms and partnerships between insects and humans. What we can confirm is the enormous significance of these relationships in shaping the nature of our planet during the coming decades. To be a "fly on the wall" is a seductive fantasy about objectivity, a human daydream of becoming the unnoticed observer of consequential events. Far more often than not, insects are the overlooked witnesses to human life. Because bugs conduct so much of their burrowing, rummaging, and scurrying between walls, in dark corners, beneath piles of leaves, and among blades of grass, we often fail to pay them heed. We are delighted to catch glimpses of bees buzzing in our gardens and butterflies alighting on our flower patches, but humans tend to listen to insects only when they pose a perceived danger.

Yet we live on a planet where they exert a remarkable degree of dominance over the vital processes of reproduction and decay, literally the beginnings and endings of all life. Such cycles, ranging from the pollination of flowering plants to the decomposition of organic matter, are fundamentally regulated by some ten quintillion six-legged creatures of countless shapes and sizes, constituting the most biodiverse group of living beings on the planet.

As I have argued throughout this book, human and insect lives are deeply entangled in ways that are not always obvious. Products generated by bugs—like shellac, silk, and cochineal—are in us, on us, and around us, linking ancient practices to modern lifestyles. It is not an overstatement to say that insects shape the ways we hear, taste, feel, and see the world. As pollinators, potential food sources, and models for laboratory research, our six-legged cousins are as central to the health of the world's food supply as they are to our intimate understanding of the human genetic code.

In 1989, a decade before the term "Anthropocene" captured the global zeitgeist, the American paleontologist Stephen Jay Gould remarked, "Don't accept the chauvinistic tradition that labels our era the age of mammals. This is the age of arthropods. They outnumber us by any criterion—by species, individuals, by prospects for evolutionary continuation. Some 80 percent of all named animal species are arthropods, the vast majority insects."* To take Gould's exhortation seriously requires us to rise above the widespread notion that planetary history is limited to human history.

---

* Since the dawn of the twenty-first century, scientists, policy makers, and pundits have referred to the current epoch as the Anthropocene. This catchword, first popularized in 2000 by the Dutch atmospheric chemist Paul Crutzen and the American biologist Eugene F. Stoermer, proclaims the unprecedented influence of humans on the planet's physical processes, asserting human responsibility for wide-ranging, and often devastating, alterations of the Earth's geology and biosphere. Certainly, the human capacity for extinguishing other species suggests the validity of this idea.

Some commentators have been perceptive enough to transcend such limitations. One of these was the reclusive but brilliant French naturalist Jean-Henri Fabre. An autodidact with no formal scientific training, Fabre became one of the nineteenth century's most widely respected entomologists. Charles Darwin and Louis Pasteur wrote letters of consultation to him, and Victor Hugo dubbed Fabre "the Homer of insects." Fabre's ten-volume *Souvenirs entomologiques* (*Entomological Memories*) is a remarkable testament to a life spent interacting with insects. At the onset of the twentieth century, Fabre remarked, "The insect does not aim at so much glory. It confines itself to showing us life in the inexhaustible variety of its manifestations; it helps us to decipher in some small measure the obscurest book of all, the book of ourselves."

Fear and awe are not as far apart on the human emotional spectrum as they so often seem. The example of Fabre's attentiveness to insect life in its manifold forms offers a bridge from dread to admiration. If we are willing to listen, our six-legged cousins give eloquent testimony about our planet's diversity and the interdependence of its human and nonhuman histories. Their lives also demonstrate how tiny fluctuations in complex systems can produce wide-ranging effects. Indeed, the flap of a butterfly's wings in Brazil can set off a tornado in Texas.

ACKNOWLEDGMENTS

Most insects are social creatures, depending on kin from their hives, nests, or anthills for safety, sustenance, and community. In similar fashion, I have relied on the convivial swarm of family members, friends, and colleagues, who helped this book take flight.

My friend and fellow environmental historian Charles C. Mann encouraged me from the outset and helped me at countless junctures. Jon Segal, my editor at Knopf, served as a constant source of inspiration and a sensible counterpoint to my meandering tendencies. My literary agent, Farley Chase, was an expert guide to the publishing world. Copy editor Bonnie Thompson made many thoughtful suggestions for how to improve my prose, and editorial assistant Erin Sellers did a marvelous job of shepherding the book to publication.

Three decades ago, at the Children's School of Science in Woods Hole, Massachusetts, Becky Lash gave me my first glimpse into the enthralling world of the entomologist. My high school biology mentor, Alison Ament, encouraged my encounters with insects in and out of the classroom. At Swarthmore College, Lillian Li helped me understand China's venerable romance with silk, and Mark Wallace inspired me to think imaginatively about human and nonhuman nature. During my years at Yale University, Jean-Christophe Agnew, Beatrice Bartlett, Bob Harms, Robert Johnston, Gil Joseph, and Jim Scott offered intellectual tenacity and capable mentorship as I crisscrossed disciplinary borders and regional boundaries. My colleagues and friends at Amherst College have supported me throughout the research and writing of this book.

Jim Boone, the Entomology Collections manager at the Bernice Pauahi Bishop Museum in Honolulu, spent hours showing me the astounding collection of insects in his care. Kirsten Carlson shared her prodigious artistic talents when designing the monarch anatomy diagram for the book, and cartographer Nick Springer did an admirable job of transferring my historical conceptions of insect commodity trade routes onto the much more legible surface of a world map.

Throughout my bug-obsessed research, Kelvin Chen generously offered his historical intuitions and entomological insights. Others who read drafts of chapters or contributed ideas along the way include: Nina Gordon, Steven Gray, Hannah Greenwald, Ken Kopp, Barbara Krauthamer, Tricia Lipton, Michael O'Connor, Dawn Peterson, Khary Polk, Elizabeth Pryor, John Soluri, Theodore Waddelow, Louis Weiss, and Ben Wurgaft. Claire Brault invited me to share my initial discoveries at Brown University's Cogut Institute for the Humanities, and Max Suechting provided a venue for my six-legged musings in his Great Books summer course at Amherst College. Likewise, Kieko Matteson gave me the opportunity to discuss my ideas about insects with her world environmental history class at the University of Hawai'i at Mānoa, and Tim McVaugh offered a similar forum in his environmental history seminar at Deerfield Academy.

Jim Gatewood, Rob Jones, Ben Madley, and Cherise Udell not only gave me detailed feedback on my work; they also provided wise counsel and boundless friendship during the challenging year when this book reached maturity.

My parents, Jerry and Lalise Melillo, deserve the "queen bee's share" of the credit for having provided unconditional love and wise guidance for more than four decades.

NOTES

INTRODUCTION

3   Despite this barrage: Zion Market Research 2018. On the intertwined development of chemical warfare and insecticides: Russell 2001. Statistics on crop damage: Culliney 2014. Insect damage as a measure of modernization: Sallam 1999.

3   From the barfly: Ammer 1989.

4   Bees, butterflies, beetles: Klein et al. 2007: 303–13.

5   From the sixteenth century onward: Bleichmar et al. 2009; Maat 2001; Drayton 2000; Grove 1995; and Brockway 1979.

6   Lorenz's talk. Lorenz 2000: 91–94. Lorenz may have been inspired by Ray Bradbury's "A Sound of Thunder": "It fell to the floor, an exquisite thing, a small thing that could upset balances and knock down a line of small dominoes and then big dominoes and then gigantic dominoes, all down the years across Time. Eckels' mind whirled. It couldn't change things. Killing one butterfly couldn't be that important! Could it?" (Bradbury 1962: 93).

6   "[There] are mysteries performed": Dillard 1974: 64.

7   *Belgica antarctica*: Usher and Edwards 1984: 19–31.

8   Following the Second World War: Vogel 2008: 667–73; Foster 2005: 3–15. I have borrowed the notion of a "synthetic planet" from Casper 2003. Conceptually, I draw upon Bruno Latour's argument that modernity is paradoxical. On the one hand it affirms that "Nature" and "Society" are distinct; on the other hand, it simultaneously depends on fusions of human and nonhuman natures that resist these strict separations (Latour 1993).

9   *Drosophila* to understand: Pandey and Nichols 2011: 411–36.

9   While entomophagy: Van Huis et al. 2013: 1.

9   "If all mankind": Wilson 1990: 6. Wilson's words are frequently misquoted and reproduced without any source citation. Thanks to Kelvin Chen for correcting this long-standing error.

10   "From the small size": Darwin 1896: 327–28.

10   "To make a prairie": Dickinson 1924: 1755.

CHAPTER 1 THE BUG IN THE SYSTEM

13 Ella Fitzgerald and the Ink Spots: Nicholson 1995: 81.

13 Hereke Imperial Carpet Manufacture: Eiland 2003: 80; Fazlẏodlu and Aslanapa 2006.

13 Six decades earlier: Myerly 1996: 68.

14 "Our lives are completely mixed up": California Academy of Sciences 2017.

14 In a related investigation: Leong et al. 2017: 6.

15 "a little straw": Washington 2003: 5.

15 "a tight-fitting suit": MacArthur 1927: 487.

15 "patiently and humbly": Hume 1956: 53.

16 "It sucked me first": Donne 1971: 58. James Joyce's 1939 novel, *Finnegans Wake*, is replete with playful couplings of "insect" and "incest." Ultimately, family members are "as entomate as intimate could pinchably be" (Joyce 1939: 417).

16 Akitushima and *kachimushi*: Kiauta 1986: 91–96; Davis 1912: xiv.

16 Japanese term *mushi* (蟲) (footnote): Evans et al. 2015: 299.

16 Manchester, England, a center of textile: Manchester City Council 2019.

16 "a statue of an ant": Flick 2006: 155.

17 "The difficult": Petty 2018: 38.

17 *Fable of the Bees:* Mandeville 1714.

18 When choosing a family emblem: Fox-Davies 2007: 260. Classicist Susan A. Stephens argues that Napoléon intentionally borrowed from the ancient Egyptians, who used a bee hieroglyph to symbolize the king of Lower Egypt (Stephens 2003: 1).

18 "If I had to choose": Sleigh 2003: 57.

19 Responding to cues: Merlin, Gegear, and Reppert 2009: 1700–04.

19 "masterpieces of integrated optics": Franceschini, Pichon, and Blanes 1992: 283.

19 Insect wings carry: Gullan and Cranston 2014; Dudley 2000.

19 The wingless insects: Norberg 1972: 247–50.

20 "What is a human being?": Laërtius 1853: 231.

20 Plato's students attempted: Bodson 1983: 3–6.

20 "In no one": Pliny 1855–57: vol. 3, 1.

20 Philemon Holland's 1601 translation: Pliny 1601.

20 "nature study": Neri 2011: xii.

20 At the other end of the century: Some historians of science also give credit to their colleagues Hans Lippershey and Jacob Metius. Bardell 2004: 78–84.

21 "flea glasses": Bradbury 1967: 68.

21 "I was plagued": Malpighi: Miall and Denny 1886: 1.

21 Merian is one: In recent years, Merian has been the subject of several biographies and a novel. Friedewald 2015; Todd 2007; Stevenson 2007.

21 In 1699: As the renowned British ecologist G. Evelyn Hutchinson wrote, Maria's voyage "was one of the first, if not actually the first, occasion on which a private individual set out across the Atlantic Ocean with the sole and express purpose of studying a problem in the New World." Hutchinson 1977: 14.

22 "Man's inhumanity": Surinam during the late seventeenth and early eighteenth centuries: Van Lier 1971 and Boxer 1965: 271–72. Brutal treatment of slaves in Surinam: Davis 2011: 925–84.

22 In stark contrast: Maria Sibylla Merian 1975: plate 2. An original edition of her *Metamorphosis insectorum Surinamensium* is in the collections of the American Museum of Natural History.

22 *Biophilia:* Wilson 1984: 1.

22 "the meeting place": J. D. Herlein quoted in Fatah-Black 2013: 8.

23 Over the coming century: Goslinga 1979: 100. For a fascinating account of a woman and her daughter who came to Surinam in 1770 as slaves, see Hoogbergen 2008. The horrifying conditions of these colonial plantations are the historical backdrop to the 2013 Dutch film *Hoe duur was de suiker* ("The Cost of Sugar"). This drama is based on the eponymous novel by Cynthia McLeod, the daughter of independent Suriname's first president, Johan Ferrier (McLeod 1987).

23 "The forest grew together": Merian 1975: plate 36. Also see plates 18, 51, and 43 (*Goliath Birdeater, Sweet Bean,* and *Marmalade Box*). Scientists have since concluded that *T. blondi* rarely consumes birds, but there is widespread agreement that these tarantulas have a cosmopolitan palate (Striffler 2005: 26–33).

24 "I almost had to pay": Rücker and Stearn 1982: 65.

24 "I created the first classification": Merian 1975: ii.

24 its author's approach: Davis 1995: 177. Art historian Janice Neri cautions that Merian's immense artistic talents and scientific innovations have sometimes been exaggerated by scholars. Merian was not the first to show the life cycle of an insect in a single image; nor was she unique among her peers in elevating insects to the realm of "high art" (Neri 2011: 141).

24 Not only did the illustrations: Etheridge 2011: 38. There are uncanny similarities between Merian and the heroine of *Mushi mezuru himegimi* (虫めづる姫君). This twelfth-century Japanese tale, whose title translates as "The Lady Who Loved Insects," focuses on the unconventional habits of a Heian court lady who befriends insects and names her attendants after her caterpillars and butterflies (Backus 1985: 41–69). Backus translated the story's title as "The Lady Who Admired Vermin."

25 Swammerdam (footnote): Cobb 2000.

25 The scientists who followed: Nation 2016; Strausfeld and Hirth 2013: 157–61.

25 Some "scout" bees: Liang et al. 2012: 1225–28.

26 *Megaponera analis:* Frank et al. 2017.

26 Insects evolved: Misof et al. 2014: 763–67.

26 The earliest fossil evidence: Hublin et al. 2017: 289–92.

26 griffinflies: Shaw 2014: 85.

26 In fact, breathing: Clapham and Karr 2012: 10927–30.

26 spiracles: Wigglesworth 1942: 194.

27 "by continuously diminishing": Chetverikov 1920: 449.

27 "A single hectare": Moffett 2010: 1. Insect population statistics: Moore 2001: 223; and Berenbaum 1995: xi.

27 "If we say 400,000": Westwood 1833: 118.

27 Estimates now suggest: Estimates of insect species range from 2.6 to 7.8 million species. Stork, McBroom, Gely, and Hamilton 2015: 7519.

27 A single family of beetles: Morris 2004: 2.

27 "An inordinate fondness": The story about Haldane is quite possibly apocryphal; see Hutchinson 1959: 146n1.

27 The petroleum fly: Kadavy et al. 1999: 1477–82.

27 Hawaiian wēkiu bug: Ashlock and Gagné 1983: 47–55.

27 In October 1988: Hoare 2009: 165.

28 In Euro-American culture: Lovejoy 1936: 236–40.

28 "It was roofed over": "Condition of Ireland," *Illustrated London News* (December 15, 1849): 394.

28 During the twentieth century: Lehane 2005.

28 "these ferocious little": Zinsser 1935: 14.

28 British naturalist: Cloudsley-Thompson 1976; McNeill 2010; Winegard 2019. For historical examples of the detrimental effects of insects on human affairs: Giesen 2011; Patterson 2009; McWilliams 2008; Sutter 2007; Lockwood 2004; and Buhs 2004.

29 435,000 people: World Health Organization 2020.

29 Meanwhile, the *Aedes aegypti:* World Health Organization 2016.

29 The Popol Vuh: Tedlock 1996: 49–50.

29 Centuries later, in North America: Lockwood 2009: 75. The bug even became known as "Shermanite," in reference to Union general William Tecumseh Sherman (Capinera 2008: vol. 3, 1766).

29 Similarly, during the Japanese occupation: Harris 1994: 79.

30 "Although these insects": Roberts 1932: 531.

30 Today, forensic entomologists: Goff 2000.

30 The residents of Enterprise, Alabama: Boissoneault 2017.

30 In the entomologist's lexicon: Schuh and Slater 1995.

30 Arabic term *baq* or *bakk:* Ibn Khallikan 1843–71: vol. 1, 234 .

30 In the United Kingdom: Schur 2013: 183.

31 "The first step": Hughes 1989: 75.

31 "First actual case": Marx 2002: 41–42; Shapiro 1987: 376–78; Yale University 2017.

31 The modern Euro-American attitude: Hoyt and Schultz 1999: 52.

32 "sweeping hatred": Anonymous 1897: 175.

32 "A world ruled by": *Modern Mechanix Magazine:* Miller 1930: 68.

32 Apocalyptic scenarios: Kinkela 2011.

32 During the California Indian genocide: Madley 2016: 325.

32 "Anti-semitism is exactly": Office of the United States Chief of Counsel for Prosecution of Axis Criminality 1946: vol. 4, 574. Also see Raffles 2007: 521–66.

32 "Something in the insects": Wheeler 1922: 386.

32 "evil arthropods": Tsutsui 2007: 237–53.

33 The insidious Xenomorph: O'Bannon 2003.

33 "We took out": *A Bug's Life* production notes (October 10, 1998), 14, quoted in Price 2009: 162.

34 The oldest known depictions: Crane 1999: 43: Toussaint-Samat 2009: 15.

34 pharaoh Nyuserre Ini: Kritsky 2015: 8–12.

34 "He that would raise": Plutarch 1875: vol. 2, 192.

35 "the honey that is found": National Archives of the United Kingdom 1225.

35 Beeswax was also a coveted: Crane 1999: 498.

35 Beekeeping evolved independently: Ransome 1937: 264. Statistics: Bianco, Alexander, and Rayson 2017: 99. In 1836, Belgian botanist Charles Morren discovered that the *Melipona* bee was the natural pollinator for vanilla orchids (*Vanilla planifola*), which are native to Mexico. Because these bees do not occur in other parts of the world, Europeans had been unable to produce vanilla beans elsewhere (Arditti, Rao, and Nair 2009: 239).

35 In the northeast corner of Australia: Fijn 2014: 41–61; Fijn and Baynes-Rock 2018: 207–16.

36 Mead, the fermented alcoholic beverage: Crane 1999: 597.

36 *Mádhu:* Turner 2008: 562; and Brothwell and Brothwell 1998: 165.

36 *Beowulf:* Enright 1996.

36 Caduveo and Bororo tribes: Lévi-Strauss 1973.

36 It was not until the emergence: Mintz 1985.

36 sugarcane and sugar beets (footnote): United States Department of Agriculture Economic Research Service 2018.

37 "Canst thou not honey me": Marston 1986: 103.

37 Winnie-the-Pooh: Milne 1926. Notably, Paddington Bear prefers marmalade. The title character in Eric Carle's *The Very Hungry Caterpillar* is an omnivore that devours any comestible in sight, be it sweet or savory.

37 Honey Nut Cheerios: Danny Hakim, "Are Honey Nut Cheerios Healthy? We Look Inside the Box," *New York Times* (November 10, 2017).

37 Similarly, another insect-derived product: Carvalho 1904: 97–101; Hahn, Malzer, Kanngiesser, and Beckhoff 2004: 234–39.

38 Unfortunately, iron gall ink: Harvey and Mahard 2014: 149; Houston 2016: 99–101; Rijksdienst voor het Cultureel Erfgoed Ministerie van Onderwijs, Cultuur en Wetenschap, 2011. Even so, the German government still used iron gall ink in certain official documents until as late as 1974.

38 Human revelations have always: A remark by Claude Lévi-Strauss, one of the founders of modern anthropology, gets to the heart of the matter: *"Les animaux sont bons à penser"* ("Animals are good to think"). The often-used English version "Animals are good to think with" is a clumsy rendering of this phrase. *"Bons à penser"* is a wording that Lévi-Strauss devised to mimic the expression *"Les animaux sont bons à manger,"* or "Animals are good to eat." "Animals are good to think" more accurately captures his intent. Lévi-Strauss 1962: 89.

CHAPTER 2 SHELLAC

39 "When you get in that groove": Berliner 1994: 349.

39 This expression from the Roaring Twenties: Almond 1995: 130; *Oxford English Dictionary* entry for "groove." https://www.oed.com/view/Entry/81733?rskey=qe5YbA &result=1#eid.

39 *The Mahabharata:* Lochtefeld 2002: 211. Very little has been written on the history of shellac. For a dated but still useful bibliography: Varshney 1970.

40 *laksha:* Yule and Burnell 1903: 499–500.

40 For thouands of years: Mohanta, Dey, and Mohanty 2012: 237–40; Buch et al. 2009: 694–703; Mukhopadhyay and Muthana 1962. Shellac cultivation also occurs in Cambodia, Indonesia, Myanmar, Thailand, Vietnam, and China. For example, the Hani ethnic group in Yunnan Province has long raised lac insects (Saint-Pierre and Bingrong 1994: 21–28). Numbers of lac insects: Negi 1996: 106. In China, Thailand, and a few other regions of Southeast Asia, *Kerria chinensis* is frequently used to produce lac (Singh 2013: 4).

40 Practitioners of Ayurveda: Dave 1950; Sarkar 2002: 224–30.

41 In 1590, the Mughal emperor: Suter 1911: 36.

41 "desks, Targets, Tables": Van Linschoten 1885: 90.

41 *"a tingere seta":* Venetian dyeing manual: Rosselli 1644: 12. For an assertion that Indian lac was being imported to Catalonia and Provence as early as 1220: Merrifield 1849: vol. 1, clxxviii.

41 "That formed on trees": vol. 2, 221. Tavernier became famous for his 1666 discovery and acquisition of a 116-carat blue diamond in India. Two years later, he sold the enormous jewel to King Louis XIV of France for the equivalent of 172,000 ounces of pure gold and a letter that granted him noble status. Eventually, gem cutters transformed the Tavernier Blue into the walnut-sized Hope Diamond, which remains one of the world's most famous precious stones (Patch 1976).

41 A seventeenth-century British treatise: Bristow 1994: 49.

42 Recent scientific analyses: Pollens 2010: 264; Steph Yin, "The Sound of Music: Secrets of the Stradivari May Be in the Wood," *New York Times* (January 3, 2017): D2.

42 These heavily fortified vessels: Chatterton 1971.

42 bound for harbors like Aden and Port Said: Arasaratnam 1986: 104.

43 *Nawāb* (footnote): *Oxford English Dictionary* entry for "nabob," https://www.oed .com/view/Entry/124762?redirectedFrom=nabob#eid.

43 "nattering nabobs": Perlstein 2008: 526.

43 "Lac enters into": Watt 1908: 1063. Also see Watt 1905: 650–52. Statistic: Misra 1928: 3.

43 Composed of more than two hundred distinct: Sainath 1996: 148–53.

44 In 1563, the Portuguese doctor: Orta 1895: vol. 2, 40. "Lakka": Watt 1905: 646. Salmasius: Watt 1908: 1055. James Kerr: Kerr and Banks 1781: 374–82. *Chermes lacca:* Roxburgh 1791: 228–35.

44 "It has been erroneously": Kirby, Spring, and Higgitt 2007: 82. Langmuir 1915: 696.

44 *Rhus verniciflua:* Webb 2000: xvii. Plant taxonomists also refer to the Chinese lacquer tree as *Toxicodendron vernicifluum.* Shellac's origins continue to be enveloped in confusion. In his 2010 book, *At Home,* renowned travel and science writer Bill Bryson mistakenly refers to lac insects as "beetles" (Bryson 2010: 156). The lac bug is actually a scale insect. At first, this might seem like a silly squabble over taxonomic details. In fact, one would have to travel back in time 372 million years to find a common ancestor shared by beetles and scale insects.

45 "friction-free capitalism": Goldsmith and Wu 2006: 57.

45 "no substitute can be found": Mackay 1861: 440.

45 "no damp air": For general information, see Dow 1927: 238; and Mussey 1981. The quote is from Stalker and Parker 1688: ii.

46 "daguerreotypes": Coe 1976.

46 "Union cases": Krainik, Krainick, and Walvoord 1988.

46 At around the same time: "William Zinsser & Company, Inc. History," http://www.funginguniverse.com/company-histories/william-zinsser-company-inc-history/.

47 Shellac entered the living rooms: Voloshin 2002: 39; and Chanan 1995: 29.

48 "the sweet tone": Thompson 1995: 138.

48 shellac-based platters: Millard 2005: 202; and Day 2000: 19.

48 In 1920 alone: Chamber of Commerce of the United States of America 1921: 19.

48 "The mixture is then put": Crandall 1924: 32–33.

49 Although aspects of this grueling procedure: Sharma 2017: 185–86.

50 "Recording is the only": Walter quoted in Kaufman and Kaufman 2003: 325. American cultural historian Michael Denning argues that the maritime circulation of 78 rpm records among the world's port cities generated the soundscape for the anti-colonial revolutions of the twentieth century. People in places as distant as Africa, Southeast Asia, and Latin America suddenly began to discover and exchange "musical vernaculars"—samba, tango, jazz—that both inspired resistance and accompanied the emergence of independent nations from the ashes of European colonialism (Denning 2015).

50 "In America I had": Stravinsky 1936: 123–24.

50 "We could never play": Dodds 1992: 71. For more on this phenomenon: Katz 2010. Origins of the expression "in the groove": Wallace 1952: 102.

51 "A hit record": Millard 2005: 203. Cosimo Matassa: Broven 1978: 106.

51 "Mr. and Mrs. Lacca": "Little Bugs and Big Business," *Popular Mechanics* 1937: 693.

51 During the Second World War: Jones 2006: 7.

51 "Owing to war conditions": *The Gramophone* (February 1943): 16. In the early 1940s, the "Big Three" record companies—RCA Victor, Columbia, and Decca—manufactured most of the records made in the United States (Dowd 2006: 208).

52 By May 1946, U.S. shellac prices: Myers 1946: 413. For examples of corporate-led recycling campaigns during wartime: Durr 2006: 361–78.

52 As late as 1953: Russ Parmenter, "Business Booming, Say Record Makers," *New York Times* (November 21, 1953): 4.

52 their ultimate demise: Shicke 1974: 120; Millard 2005: 204; Frith 2004: 277; and Granata 2002: 8.

53 "I'm not recommending": Barack Obama: White House, Office of the Press Secretary 2010.

53 "juke house": Mitchell Landsberg, "Oldies but Goodies: Jukebox Turns 100 and Brings Back Memories of Old-Time Rock 'n' Roll," *Los Angeles Times* (November 19, 1989): 2.

54 "After shellacking the alleys": La Point 2012: 114.

CHAPTER 3 **SILK**

56 "Light as a feather": Illica, Giacosa, Elkin, and Puccini 1906: 9.

56 Moths and butterflies both belong (footnote): Scott 1986: 95.

56 "The silkworm-moth eyebrow": Hearn 1899: 59.

56 "Her forehead cicada-like": "Shuo Ren," poem 57 in *The Book of Odes* (*Shijing*) Ward 2008: 36.

56 *éméi* (蛾眉): Kyo 2012: 15.

56 The moth's crescent-shaped antennae: Hargett 2006: 26.

58 In the 1850s, the Irish travel writer: Crawford 1859: 187.

58 One such colonist, Priscilla Jacobs: Arrington 1978: 381.

58 "The eggs when on paper": Wood 2002: 32. While wild silk has been gathered since at least the third millennium B.C.E., sericulture developed in China by around 2700 B.C.E. (Whitfield 2018: 191–92; and Datta and Nanavaty 2005: 113).

59 A mature silkworm's two salivary glands: Sue Kayton, correspondence with author; Sun et al. 2012: 483–96; and Millward 2013.

59 ahimsa silk (footnote): Gajanan 2009: 421–24.

61 According to Confucius: Kuhn 1984: 213–45. Confucius lived c. 551–479 B.C.E.

61 "Either way": Eugenides 2002: 63.

61 "Folded just so": John McPhee, "Silk Parachute," *New Yorker* (May 12, 1997): 108.

61 The funerary garment: this silk gauze was discovered in 1984 at Qingtai Village in Henan Province (Schoeser 2007: 17).

62 "Now the [Xiongnu]": Yu 1967: 11.

62 The Romans first encountered: Bulnois 1966: 10; and Hopkirk 1980: 20. Perhaps the account of the silk banners is apocryphal. No mention of it appears in either Plutarch's *Life of Crassus* or Cassius Dio's *Roman History* (Plutarch 1916; and Cassius 1914–27: vol. 3, book 40, 437–47).

63 "prepared cords": Arrizabalaga y Prado 2010: 330. The quote is from Anonymous, *The Scriptores Historiae Augustae*, vol. 2, 171.

63 "The use of silk": Ammianus Marcellinus quoted in Hudson 1931: 77. The Byzantine Empire was the continuation of the Eastern Roman Empire. Pliny (the Elder) 1855–57: vol. 2, 36–37 ("so famous"). Also see Vainker 2004: 6.

64 Conjecture aside: Fong 2004: 6; and Bozhong 1996: 99–107.

64 "Throughout the village": Hammers 1998: 197. In 1145, Lou Shou penned an influential set of forty-five verses on rice cultivation and silkworm rearing. These poems, collected in *Pictures of Tilling and Weaving* (*Gengzhi tu* 耕織圖)—published a century later with an accompanying array of woodblock prints—became iconographic for China's imperial officials.

64 Its body weight will increase: Sue Kayton, correspondence with author.

65 Aspiring silk makers the world over: As Chinese historian Lillian M. Li noted, "Mulberry trees required a great investment of time as well as land. [Chinese] sericultural manuals pointed out that if one got discouraged before the six or seven years necessary for a tree to mature had elapsed, all the previous effort could be wasted" (Li 1981: 142).

65 Sometime during the first century C.E.: Hill 2009: 466–67.

66 "The secret": McLaughlin 2016: 11.

66 "and in the manner": Procopius 1928: 229–31.

66 During the Second Crusade: Muthesius 2003: vol. 1, 325–54.

67 *die Seidenstraßen*: Frankopan 2017: xvi. Richthofen's travels: Elisseeff 1998: 1–2. Ferdinand was the uncle of Manfred von Richthofen, the World War I flying ace known popularly as "the Red Baron."

67 *Pax Mongolica*: Beckwith 2009: 183.

67 "a maiden": Lewis and Morton 2004: 121.

67 By 1257, more than a decade: Lopez 1952: 73.

67 *china poblana*: Seijas 2014: 27.

67 Sixteenth-century Mexican dressmakers: Parry 1966: 132.

68 "Above all": Schurz 1939: 32.

69 *qipao* (旗袍) (footnote): Ma 1999: 41.

69 The most notorious: Davis, 2001.

69 When the Huaynaputina volcano: Verosub and Lippman 2008: 141–48; and Thirsk 1997: 120.

69 "By all likelihood": Peck 2005: 87.

70 In 1685, Louis XIV: Hertz 1909: 710; and Scoville 1952: 300.

70 "By the dwellings": Smith 1910: vol. 1, 56. For comprehensive treatments of sericulture history in Virginia and colonial North America, see Hatch 1957 and Klose 1963. U.S. silk production did not expand markedly until after the Civil War, when British silk entrepreneurs set up shop in towns like Paterson, New Jersey.

70 "a kind of El Dorado": Morgan 1962: 147.

71 "larding the lean earth": Stoll 2002.

71 "bring as food": Miller and Gleason 1994: 38.

71 "The southerners": Xue 2005: 60.

71 For example, in precolonial East Africa: Oliver 1975–86: vol. 3, 640. For more examples of manure fertilizers, see Mather and Hart 1956: 25–38.

71 "A great city": Hugo 1915: 84.

72 "mulberry embankment and fish pond" system: Marks 1997: 119. Alvin So argues that as much as one-fourth of China's total silk exports before 1840 came from

Guandong (So 1986: 81n2). China's other major silk-producing region was the Yangzi River Delta. Historian Kenneth Pomeranz compares silk production in these two regions during 1750 (Pomeranz 2000: 327–38).

72 Governor-General Xiliang: Yimin 1995: 143.

73 "Next to brass": Adkins and Adkins 2008: 161–63 (emphasis added).

73 "A well-tied [silk] tie": Wilde 1894: 90. Although Wilde's satirical intentions are on display here, he was quite a dandy in his own right. As he remarked in the second chapter of *The Picture of Dorian Gray,* "It is only shallow people who do not judge by appearances" (Wilde 1891: 34). It takes 220 pounds of mulberry leaves to yield 25 silkworm cocoons. It takes approximately 3,000 cocoons to make one pound of silk. Sue Kayton notes, "It requires 1700 to 2000 cocoons to make one silk dress (or about 1,000 cocoons for a silk shirt)" (correspondence with author).

74 "No industry": Challamel 1882: 98. The quote is from Beauquais 1886: 248.

75 "For over six years": Trouvelot 1867: 32.

75 "A cluster of tiny eggs": Spear 2005: 23. Also see Liebhold, Mastro, and Schaefer 1989: 20–22. Unlike tent caterpillars (genus *Malacosoma*), for which they are frequently mistaken, gypsy moths do not build "tents" or nests in the trees they infest (UMass Extension 2015: 1).

75 Abandoning the insect world: Trouvelot 1882. Details about these: Rees 1999: 80.

76 "Like other members": Wood 2002: 33.

76 "They were called": Henderson 1958: 129. A sense of loss connected humans to silkworms in other ways. In 99 B.C.E., the renowned Chinese historian Sima Qian (c. 145–c. 86 B.C.E.) was castrated for outspokenly defending a military officer who had lost a campaign against nomadic tribal armies on the Han Empire's northern frontier. Sima referred to his ordeal as going down to the "silkworm chamber" (蛾眉; *cán shì*) (Durrant 1995: 150).

77 "reveal the chic": Glickman 2005: 573.

77 The editors of *DuPont Magazine:* Ndiaye 2007: 101.

77 "It is an entirely new": *Fortune* magazine: Hermes 1996: xiv.

77 "Biological materials": Pérez-Rigueiro, Viney, Llorca, and Elices 1998: 2439. Regarding new applications of silk, see Leal-Egaña and Scheibel 2010. Spider silk is even stronger than silkworm silk, but it has proved extremely difficult to produce in large quantities. Unlike silkworms, spiders are not social insects. They don't tend to "play well together," so it is hard to raise them in large enough numbers to scale up spider silk production (Shao and Vollrath 2002: 741).

77 "Did you know": Anonymous 1874: 165.

77 "Silk drives engineers": Lawry 2006: 18.

CHAPTER 4 COCHINEAL

78 French biopirate: The first use of the term "biopiracy" was in a Rural Advancement Foundation International (RAFI) newsletter from 1993: RAFI, "Bio-Piracy: The Story of Natural Colored Cottons of the Americas," *RAFI Communiqué* (Novem-

ber 1, 1993): 1–7. The concept describes the theft of biological resources, often from less affluent countries or marginalized peoples.

78 "new Argonaut": Thiéry de Menonville 1787: vol. 1, cvi. All translations from the French are mine.

79 two types of cochineal (footnote): "Cochineal." *PubChem* (U.S. National Library of Medicine), https://pubchem.ncbi.nlm.nih.gov/compound/Cochineal. To correctly pronounce "cochineal," keep in mind the humorous image "coach an eel."

79 Much of this pricey dye: Phipps 2010: 33.

79 "Its price": Thiéry de Menonville 1787: vol. 1, ciii.

80 "tired, disgusted": Ibid.: vol. 1, 4.

80 In 1791: James 1938.

80 He decided to pose: Thiéry de Menonville 1787: vol. 1, 76.

80 "a quantity of vials": Ibid.: vol. 1, 5–6.

80 Arriving in Havana: Ibid., vol. 1, 105.

81 "a river of silver": Galeano 1987: vol. 2, 94. The population of Veracruz in 1791 was roughly eight thousand (Knaut 1997: 622).

81 Thiéry disembarked at the Veracruz docks: Thiéry de Menonville 1787: vol. 1, 59.

82 "brusque tone" and "foul language": Ibid.: vol. 1, 57.

82 "Such a discovery": Ibid.: vol. 1, 60. The city's residents may also have been pleased because of the economic implications of Thiéry's discovery. Jalap root was one of the drug plants the Spanish exported from the Americas to Europe starting at the beginning of the seventeenth century (Williams 1970: 399–401).

82 "golden fleece": Thiéry de Menonville 1787: vol. 1, 138.

83 "Without a coat": Ibid.: vol. 1, 126–27.

83 "nearly naked" and "Her charms": Ibid.: vol. 1, 69–70.

84 "He responded," "the tiny insects," and "the true purple": Ibid.: vol. 1, 114.

84 The Frenchman and his treasured: Ibid.: vol. 1, 145.

84 "two viceroys": Ibid.: vol. 1, 184.

84 "I added": Ibid.: vol. 1, 208–09.

85 *gens de couleur*: Ibid.: vol. 1, civ. The annuity of six thousand livres was two and a half times the salary paid to the colony's royal physicians. Cap Français is now known as Cap Haïtien; this city on the north coast of Haiti was nicknamed "the Paris of the Antilles" because of its wealth and cultural sophistication.

85 René-Nicolas Joubert de la Motte: Moreau de Saint Méry 1798: vol. 2, 367–68.

85 A private colonist: McClellan 2010: 156.

85 The discovery of carmine-infused: Saltzman 1986: 27–39.

86 The Aztecs, who inherited: Rodríguez and Niemeyer 2001: 76. Historically, Mesoamerica was a cultural zone that extended from central Mexico southward through Belize, Guatemala, El Salvador, Honduras, Nicaragua, and northern Costa Rica.

86 *nocheztli*: Ferrer 2007: 58. Other insect commodities that rely upon domesticated host plants include the maguey worm (*Aegiale hesperiaris*), grown on *Agave tequilana* leaves for the tequila industry (Staller 2010: 39–40).

86 The Codex Mendoza: Folio 42v of the sixteenth-century Codex Mendoza lists trib-

ute items from around the Aztec confederacy, including sacks of cochineal (Berdan and Anawalt 1997: 90–91). The cochineal bug was one of only five animals the Aztecs domesticated. The others were the turkey (*Meleagris gallopavo*), the Muscovy duck (*Cairina moschata*), the dog (*Canis lupus familiaris*), and the now-extinct species of North American honeybee (*Apis nearctica*) (Conlin 2009: 16–17).

86 "They have colors": Cortés 1962: 88.

86 Despite widespread misconceptions: Grimaldi and Engel 2005: 301.

87 The sought-after red dyestuff: Eisner, Nowicki, Goetz, and Meinwald 1980: 1039.

87 *"la grandeza"*: Cobo 1890–95: vol. 1, 445.

87 Like lac cultivation: Chávez Santiago and Meneses Lozano 2010: 2; and Donkin 1977: 15.

88 In preparation for cochineal production: Baskes 2000: 129–30; and Dahlgren de Jordán 1961: 387–99. Packed the dye into leather sacks: Downham and Collins 2000: 12.

88 Spanish colonizers did little: Pomeranz and Topik 2006:115.

88 "Everyone does nothing": Mills and Taylor 2006: 91. Thanks to John Soluri for bringing this source to my attention.

88 "[Tlaxcala] takes in": Donkin 1977: 26.

89 twenty-five to thirty thousand residents: Baskes 2005: 192.

89 "This *grana*": Sahagún 1829–30: vol. 3, 287. The *Historia general de las cosas de la Nueva España* (*General History of the Things of New Spain*), also known as the Florentine Codex, is the most complete work on indigenous colorants in New Spain. Fray Bernardino de Sahagún produced the *Historia general* in Mexico City from 1575 to 1580. However, it was not published until the early 1800s. Because the three volumes (divided into twelve books) are held at the Laurentian Library in Florence, Italy, they are collectively referred to as the Florentine Codex. This work contains more than two thousand illustrations drawn by native artists and remains one of the most comprehensive accounts of indigenous life at the time of the Spanish conquest. The other useful source for sixteenth-century Mexican dyestuffs is the Badianus Manuscript, also referred to as the *Libellus de Medicinalibus Indorum Herbis* or the Codex Barberini. Two Aztec scribes compiled it in 1552 at the Colegio de Santa Cruz in Tlatelolco, Mexico.

89 The arrival of Latin American cochineal: The concept of the "Columbian Exchange" originates with Crosby 1972.

89 Upwards of fifty-six million people: Black 1992: 1739.

90 "reknit the seams of Pangaea": Mann 2011: 6.

90 "If Columbus": Voltaire 1876–78: vol. 8, 379. Columbus and syphilis to Europe: Choi 2011.

90 Long before the Genoese navigator: Greenfield 2005: 1–2.

90 Prior to the arrival of cochineal: Leggett 1944: 69–82; and Donkin 1977: 7.

90 However, once Mexican cochineal: Hofenk–De Graaff 1983: 75; and Lee 1951: 206.

90 In 1599, the respected: Gómez de Cervantes 1944: 163–64; and Brading 1971: 96.

90 Indeed, by the eighteenth century: Salazar 1982.

90   Cochineal invigorated: Elizabeth Malkin, "An Insect's Colorful Gift, Treasured by Kings and Artists," *New York Times* (November 28, 2017): C2.

91   "Macready stumbled off": Francis 2003: 73. Another account of the incident: Schreiber 2006: 132.

91   As with shellac and silk: Petty 1702: 796–97; and Cowan 1865: 261.

91   "The Cochinilla is not": Hakluyt 1903: vol. 9, 358.

91   "Concerning Cochineel": Donkin 1977: 3. In *The Wheels of Commerce*, Fernand Braudel recounted the failed attempt by the Dutch banking house Hope & Co. to achieve a monopoly over the cochineal trade in 1787 (Braudel 1992: vol. 2, 421–23).

92   British cultivators enjoyed: Chávez-Moreno, Tacante, and Casas 2009: 3347; and Hamilton 1807: vol. 3, 399–400.

92   cactoblastis moths: State of Queensland, Department of Agriculture and Fisheries 2016. In the 1850s, German physician and botanist William Hillebrand developed an elaborate plan to raise the dye-producing insect at Foster Botanical Garden in Honolulu. In large part, confusions over the correct host plants for domesticated cochineal thwarted these attempts to start a Hawaiian cochineal industry. William Hillebrand, "The Cochineal," *Commercial Advertiser* (August 20, 1857): 1.

92   "The red coats": Anonymous, "Good Lac," *Anglo-American Magazine*, vol. 3 (1853), 297. Concurrently, Emily Dickinson was beginning her prolific writing career. Several of her poems featured cochineal, including "A Route of Evanescence," in which she describes a hummingbird as "A rush of cochineal" (Dickinson 2003: 91).

92   Long before the Spanish: Phipps 2010: 10. When a soluble dye (like cochineal) is precipitated with a metallic salt, it is called a "lake."

93   indigo (*Indigofera tinctoria*): McCreery 2006: 53–75; and Balfour-Paul 1998.

93   campeche wood: Contreras Sánchez 1987: 49–74. Among the reasons that brazilwood never became a major alternative for European dyers was the rapidity with which colonists extracted the tree from Brazil's coastal ecosystems. By 1605, the Portuguese Crown had already issued legislation to protect surrounding forests from rampant cutting. These attempts at conservation proved futile (Frickman Young 2003: 105).

93   "The insect was not": Edwardes 1888: 50. Other sources place the introduction of cochineal from Mexico to the Canary Islands via Cádiz at 1820 (Gonzáez Lemus 2001: 178).

93   Extracted from coal tar: Travis 2007: 2.

93   "But recent discoveries": Romero 1898: 53. Rise of aniline dye production and the corresponding decline in Mexico's cochineal exports: Coll-Hurtado 1998: 81.

93   As a Virginia newspaper: "COCHINEAL IS NEAR END: Soon to Become Thing of History Like Tyrian Purple of Antiquity," *Alexandria Gazette* (January 6, 1912): 3.

94   In 1990, the FDA: Hamowy 2007: 188.

94   The nontoxicity: Schul 2000: 1–10; Wrolstad and Culver 2012: 59–77; and Jane Zhang, "Is There a Bug in Your Juice? New Food Labels Might Say," *Wall Street Journal* (January 27, 2006): B1.

94   "Carminic acid is one": Wild Colors from Nature website 2013.

94 In April 2012, Starbucks: Karin Klein, "Starbucks Is Getting the Bugs Out," *Los Angeles Times* (April 23, 2012). Cases of adverse health reactions to cochineal: Cindy Skrzycki, "Allergy Fears Tinge Debate on Bug-Dye Rule," *Washington Post* (May 9, 2006): D1.

94 While Mexican farmers: Rodríguez and Niemeyer 2001: 78; Carlos Rodríguez and Pascual 2004: 243–52. The Canary Islands archipelago has also reinvigorated its cochineal production: Desiree Martin, "Spanish Islands Launch a Cochineal Comeback," *Taipei Times* (September 18, 2011): 11.

94 "The Red That Colored the World": Padilla and Anderson 2015.

95 Contemporary artists continue: *Red Room* (http://elenaosterwalder-atelier.com/#/amati-installation/). *Amate* paper had been made by indigenous residents of Mexico long before the Spanish invasion. In fact, the Spanish word comes from the Nahuatl (Aztec) word *āmatl*. During colonization, the Spanish banned amate production because of the paper's association with indigenous religious practices (López Binnqüist 2003: 115).

CHAPTER 5 RESURGENCE AND RESILIENCE

96 "I must tell you": Jaffe 1976: 131. These results appeared in Wöhler 1828: 253–56. For a summary, see Kinne-Saffran and Kinne 1999: 290–94. For a compound to be organic, it must contain both carbon and hydrogen.

96 "The substances which": Wilkes and Adlard 1810: vol. 4, 167 (emphasis added).

97 "The prime consequences": Le Corbusier 1927: 232.

97 "A house is a machine": Le Corbusier 1927: 95. Other prominent modernist architects of the day included Walter Gropius, Alvar Aalto, Ludwig Mies van der Rohe, and Frank Lloyd Wright.

97 The 1950s marked: Killeffer 1943: 1140–45.

97 "In the future": Adams 1952: 163. In his 1989 book, *The End of Nature*, environmentalist Bill McKibben described Adams as "a glib Dacron worshiper" (McKibben 2006: 70).

97 "The time has come": Rosin and Eastman 1953: 7. In the early 1900s, Eastman had been a prominent political activist in New York and a leading patron of the Harlem Renaissance. By the 1950s, he had abandoned his early support for socialism and radical causes. Eastman became a staunch anti-Communist and joined the classical liberal Mont Pelerin Society, founded by Friedrich von Hayek, Ludwig von Mises, and others.

97 A few years later: Carroll Kilpatrick, "Economic Problems in Synthetics Cited: One in Series of Reports," *Washington Post* (September 20, 1959): B12.

98 "In fact, it could be": Anonymous 2006: 53; Mannheim 2002: 52–54; Cooper 1989: 76.

99 "Despite having names": Barthes 1972: 97. French ambivalence toward plastic and other synthetics: Smith 2007: 135–51.

99 Bakelite (footnote): "New Chemical Substance: Baekelite Is Said to Have the Properties of Amber, Carbon, and Celluloid," *New York Times* (February 6, 1909); and Meikle 1995: 31–62. History of plastics: Fenichell 1996.

99 "how the typical": Avila 2004: 16.

100 In an era dominated: Bird 1999: 23. In the 1970s, Monsanto employed a similar catchphrase: "Without chemicals, life itself would be impossible" (Noble 1979: 24). DuPont later shortened the slogan to "Better Living Through Chemistry," which the company used until 1982. From 1982 to 1999 it was "Better Things for Better Living."

100 At precisely the same moment: Firn 2010: 127–39.

100 The study of the harmful effects: Wexler 2018: 437.

100 The Ebers Papyrus: Nunn 1996: 30–34.

100 *pharmakon* and *du:* Liu 2015: 89–97.

100 *Archiv für Toxikologie:* The journal's original name was *Sammlung von Vergiftungsfällen,* which translates to "Collection of Poisoning Cases."

100 Ransdell Act and the NIH: Davis 2008: 674–83.

101 They included London's: Wise, 1968; Logan 1953: 336–38.

101 Minimata, Japan: George 2001.

101 thalidomide: Stephens and Brynner 2001; and McFadyen 1976: 79–93.

101 "Toxicity is suddenly": Coon and Maynard 1960: 19.

101 "500 new chemicals": Carson 1962: 7. Six months before *Silent Spring* hit bookstores, Alfred A. Knopf published Murray Bookchin's *Our Synthetic Environment.* Bookchin's pioneering volume, released under the pseudonym Lewis Herber and introduced by eminent soil scientist William A. Albrecht, offered an unprecedented clarion call about the dangers of pesticide use to the environment and human population of the United States (Herber 1962).

101 "Why a spinster" and "probably a Communist": Lear 1989: 429. A description of the attacks appears in that book in chapter 8.

102 In the mid-1970s: Gibbs 1982. Health consequences. Janerich et al. 1981: 1404–07.

102 Of similar effect: Chouhan 1994. Health consequences: Mehta et al. 1990: 2781–87; and Taylor 2014.

102 Government statistics: Mitchell 2002: 21.

103 "debates over chemicals": Nash 2008: 651.

103 The ensuing cultural backlash: Colborn, Dumanoski, and Myers 1996. The major glands of the endocrine system are the hypothalamus, pituitary, thyroid, parathyroids, adrenals, pineal body, and the reproductive organs (ovaries and testes).

103 "chemical regime": Murphy 2008: 697. Also: Davis 1998; and Wargo 1998.

103 In 2018, medical researchers: Wang et al. 2018.

103 "ghost acreage": Michigan State University food scientist Georg Borgström first developed the concept of "ghost acreage" in 1965. Fifteen years later, environmental sociologist William Catton Jr. extended Borgström's concept to include past "fossil acreage," or "imports of energy from prehistoric sources" (Borgström 1965; and Catton 1980: 41). Contradictions that arise when substituting labor and capital for natural resources: Ayres 2007: 115–28; and Hornborg 2001: 32.

104 Often carried out: Beattie 2011; Prudham 2005; and Scott 1998.

104 In 1950, children: Deshpande 2002: 227. Delaney Clause: United States Statutes: 21 USC. 348(c)(3)(A) (199).

104 "At least for now": Renault 2019. Also: Mozzarelli and Bettati 2011: xxiv; and Squires 2002: 1002–05.

105 Natural latex: Finlay 2009. Statistic: Charles C. Mann, personal communication. For a 1955 prediction that natural latex would soon be replaced by synthetic latex: Solo 1955: 55–64.

105 When it comes to vanilla: Firn 2010: 80; Reineccius 2006: 250; and Kahane et al. 2008: 23–29.

105 "silk filaments": Holland, Vollrath, Ryan, and Mykhaylvk 2012: 105 and 108.

105 The global trade in textiles made: BusinessWire 2017.

105 The unit price: Schoeser 2007: 180.

105 Contemporary sericulture shares: Datta and Nanavaty 2005: 10–11; Geetha and Indira 2011: 89–102; Earth et al. 2008: 43–66; and Rani 2006.

106 For example, Indian silk weavers: Islam and Hossain 2006: 158.

106 Their tightly woven mesh: Huq et al. 2010: 1–5; and Abigail Zuger, "Folding Saris to Filter Cholera Contaminated Water," *New York Times* (September 27, 2011): D7. Derived from the Sanskrit word *sati,* which translates as "a strip of cloth," "sari" can also be spelled "saree."

106 "My mom dressed me": Ginsberg 2007.

106 Products developed in Thailand: Francis Childs, "Rub Your Face with Silkworm Cocoons to Wipe Away Wrinkles. It Sounds Bizarre—but It Works," *Daily Mail* (November 16, 2014).

107 "Shellac's main appeal": Owen 2006: 226. Also: Allen 1994: 36–46.

107 "something profoundly tactile": Petrusich 2014: 239.

107 Dedicated anglers: Wright 2015: 30–31; and Keene 1891: 71.

107 Because of its approval: Altman 1998: 98.

108 Likewise, the pharmaceutical industry: Ma, Qiu, Fu, and Ni 2017: 53401–06. Other uses: Stummer et al. 2010: 1312–20; Le Coz et al. 2002: 149–52; Al-Hayani et al. 2011: 1561–67; and Hoang-Dao et al. 2009: 124–31.

108 The magnitude of disposed: Alana Semuels, "The World Has an E-Waste Problem," *Time* (May 23, 2019).

108 "tsunami of e-waste": United Nations News Center 2015.

108 Researchers in Austria: Feig, Tran, and Bao 2018: 337–48; and Irimia-Vladu et al. 2013: 1473–76.

108 Although Cambodia, Indonesia: Ghosh 2015: 105; India Brand Equity Foundation 2018. Over 90 percent of Indian lac comes from six states: Jharkhand, Bihar, Madhya Pradesh, West Bengal, Maharashtra, and Odisha. Numbers employed in lac cultivation: Bhardwaj and Pandey 1999–2000: vol. 6, 229; and Sharma, Jaiswal, and Kumar 2006: 894.

108 In Jharkhand and elsewhere: Sarin 1999: 238; Kameswari 2004: 169; and Sharma and Kumar 2003: 80. NTFPs include almost any product gathered from forests that is

not tree wood. These include: fruits and nuts, vegetables, fish and game, medicinal plants, resins, essences, sustainably harvested barks like cork or cinnamon, and fibers such as bamboo, rattans, and other palms and grasses.

109 Today, Peru: Müller-Maatsch and Gras 2016: 385–428.

109 The global market for cochineal: Grand View Research 2018.

109 Buried deep within: Imbarex Natural Colors & Ingredients 2019.

109 This belief began: Schweid 1999: 160.

110 "the cockroach, a venerable": "The Insects Shall Inherit the Earth," *Nation* (March 3, 1962): 187. Portions of this article (not direct quotes from Glass's speech) reappeared in Milton Bracker, "Atomic War Held Threat to Nature; Only Insects and Bacteria Will Survive, Scientist Says," *New York Times* (June 17, 1962): 46.

110 "The winner": *New York Times* (July 22, 1965): 13. Cockroach survivor myth: Robert W. Stock, "It's Always the Year of the Roach," *New York Times Magazine* (January 21, 1968): 34–39.

110 "For sheer survival": Catherine Breslin, "Coping with the Cockroach," *New York Magazine* (March 9, 1970): 57.

111 "The idea that": Grupen and Rodgers 2016: 86.

111 During the 2008 season: *MythBusters* 2008. The Swiss scientific illustrator Cornelia Hesse-Honegger has devoted her career to documenting the mutations that severe radiation exposure causes in insects. Her intricate watercolors depict the bodies of deformed bugs from the fallout areas of Chernobyl and other nuclear facilities around the world (https://atomicphotographers.com/cornelia-hesse-honegger).

111 "living fossils": Datta 2010: 55.

111 "Of no other type": Bell, Roth, and Nalepa 2007: 37.

111 *M. rhinoceros* and *A. fungicola*: Bell, Roth, and Nalepa 2007: 6–7. To date, entomologists have identified forty-seven species of leaf-cutter ants, belonging to the two genera, *Atta* and *Acromyrmex* (Speight, Watt, and Hunter 1999: 156).

112 "a certaine India Bug": Smith 1624: vol. 5, 171.

112 "The cockroaches that had crawled": Calonius 2006: 110–11. The United States officially outlawed the importation of slaves in 1808, but the practice continued through clandestine means. The *Wanderer* was probably the penultimate ship to ply this slave route. The *Clotilda*, which carried 110 slaves from Dahomey in 1860, is the last known ship to bring African slaves to the States (Diouf 2007).

112 "appears to have": Miall and Denny 1886: 41.

112 On July 18, 1989: Song et al. 2003: 243.

112 "In Kafka's correspondence": Kafka 2014: 121.

113 Archy's jovial disposition: Beezley and Ranking 2017: 128; and Rodríguez Marín 1883: 328–75. Some versions of "La Cucaracha" deride Huerta for his marijuana-smoking habits. The cockroach cannot walk *"porque le falta marihuana que fumar"* ("because it lacks marijuana to smoke").

114 In her haunting memoir: Mukasonga 2016.

114 On April 17, 2015: Katie Hopkins, "Rescue Boats? I'd Use Gunships to Stop Migrants!" *Sun* (April 17, 2015): 11.

114 Hopkins penned her hateful screed: Patrick Kingsley, Alessandra Bonomolo, and Stephanie Kirchgaessner, "700 Migrants Feared Dead in Mediterranean Shipwreck," *Guardian* (April 19, 2015).

114 As the researchers discovered: Fardisi, Gondhalekar, Ashbrook, and Scharf 2019.

115 In 2007, a team of scientists: Inward, Beccaloni, and Eggleton 2007: 331–35.

115 "That was the year": Stephen Vincent Benét, "Metropolitan Nightmare," *New Yorker* (July 1, 1933): 15. Benét achieved fame as the author of the long poem *John Brown's Body*, which won the Pulitzer Prize in 1929.

115 Insects, like so many: Kolbert 2014; and Brooke Jarvis, "The Insect Apocalypse Is Here," *New York Times Magazine* (November 27, 2018): 40–45, 67, and 69.

116 The first global: Sánchez-Bayo and Wyckhuys 2019: 8–27; and Hallman et al. 2017.

116 "Extinction is not mere death": MacKinnon 2013: 37.

CHAPTER 6 NOBEL FLIES

119 *Ceratitis capitata* (footnote): Thomas et al. 2019: 1. In Australia, South Africa, and parts of South America, Mediterranean fruit flies are considered economic pests because of the damage they inflict on fruit crops. *Drosophila melanogaster* used to be called the "vinegar fly" or the "pomace fly." Scientists published the complete genome sequence of *Drosophila* in 2000 (Adams 2000: 2185–95).

120 Thomas Hunt Morgan: Morgan was not the first scientist to work with *Drosophila melanogaster*. In 1901, geneticist William Ernest Castle and his students at Harvard University began using fruit flies as model organisms. Castle's pioneering work, *Genetics and Eugenics* (1916), laid the foundation for the study of heredity in animals (Snell and Reed 1993: 751–53). Today, researchers also raise *Drosophila* on diets of cornmeal, sugar, and yeast.

120 As a child: Shine and Wrobel 1976: 1–30. On August 20, 1866, the United States Civil War formally concluded when President Andrew Johnson signed "Proclamation 157—Declaring That Peace, Order, Tranquility, and Civil Authority Now Exist in and Throughout the Whole of the United States of America" (https://www.loc.gov /item/rbpe.23600100/).

120 Charlton Hunt Morgan: Shine and Wrobel 1976: 5.

121 A former whaling village: "History of the MBL," https://www.mbl.edu/history -of-the-mbl/.

122 Sampson, who took: Keenan 1983: 867–76.

122 "Of course I expected": Shine and Wrobel 1976: 112.

123 "appeared at the door": Ibid.: 2.

123 Summer School Club at Woods Hole (footnote): Keenan 1983: 867–76.

123 "I had my first prolonged": Carson 1998: 148.

124 "How's the fly?" and "How's the baby?": Cauchi 2014: 19.

124 Previous research on inheritance: Morgan 1910: 120–22.

124 Over the course: Fisher and De Beer 1947: 451–66.

125 "Time flies like": Kanfer 2000: 5.

125 "a biological breeder reactor": Kohler 1994: 47.

125 "An *E. coli* bacerium": Weiner 1999: 9.

126 During the warmer months: Mawer 2006: 3.

127 "friendly of countenance": Gustafsson 1969: 240.

127 without its controversies (footnote): Fisher 1936: 115–26; and Pires and Branco 2010: 545–65.

128 "Not a soul": Henig 2000: 169.

128 "Investigations which before": Bateson 1913: 2.

128 "For example, if you mix": Sturtevant 2001: 3.

129 "I come at this Christmas": Bateson 1928: 392.

130 "bred from a jar": Lintner 1882: 216.

130 Woods Hole grocery store: Sturtevant 2001: 3.

130 *Drosophila* flies are also: Keller 2007: R77-R81.

130 "Everywhere [academic] chairs": MacBride 1937: 348.

131 "At times, it may even": Greenspan 1997: 1. At the beginning of the twentieth century, the German sociologist Max Weber called the replacement of superstition with rationality "the disenchantment of the world" (Weber 1946: 148). Weber adopted this phrase from the German dramatist, poet, and philosopher Friedrich Schiller.

131 "Chamber of Secrets": Jaubert, Mereau, Antoniewski, and Tagu 2007: 1211–20.

131 "Wizardry and Artistry": Ejsmont and Hassan 2014: 385–414.

131 "Am not I": Blake 1977: 124.

132 "Five hundred generations" : Lindsay 2002: 30–31.

132 Morgan's protégés used: Wangler, Yamamoto, and Bellen 2015: 639–53.

132 Three American researchers: Nobel Assembly at Karolinska Institutet 2017.

133 During the months before: Siegel 2009: 5–6. In all likelihood, Palin's aides had fed her the talking point after becoming confused about a small U.S. grant to fund research on the olive fruit fly, *Bactrocera oleae*.

133 Palin's ignorance: Andrews et al. 2008: 3839–48.

133 "The nearest gnat": Whitman 2007: 65.

CHAPTER 7 **LORDS OF THE FLORAL**

134 The package had begun: "Biddulph Grange Is a Horticultural Disneyland," *Independent* (September 24, 2006).

134 "Good Heavens": Darwin 1862, "Charles Darwin to Joseph Dalton Hooker." The episode is recounted in Arditti, Elliott, Kitching, and Wasserthal 2012: 403–32. Joseph Dalton Hooker (1817–1911) was an eminent botanist known for his scientific globe-trotting and his contributions to plant geography. He was among Darwin's most devoted supporters. Hooker served as director of the British Royal Botanic Gardens, Kew, from 1865 to 1885 (Endersby 2008).

134 "In Madagascar": Darwin 1896: 44.

135 "we find nothing": Campbell 1867: 44.

136 "That such a moth exists": Wallace 1867: 471–88.

136 That year, the zoologist: Kritsky 1991: 206–10. Darwin died in 1882, but Wallace lived until 1913.

136 That year, Professor Philip Devries: Netz and Renner 2017: 471.

137 Some angiosperms: Ollerton, Winfree, and Tarrant 2011: 321–26.

137 "Lords of the floral": Darwin 1908: 303.

137 This coevolutionary dance: Holden 2006: 397.

137 Annually, upwards of half: Food and Agriculture Organization of the United Nations 2016.

138 The tiny, bright blossoms: Weiss 1991: 227–29.

138 "stopping here and there": Oliver 1986: 50.

138 As the bees repeat: Woodgate, Makinson, Lim, Reynolds, and Chittka 2017: 1–15.

138 he became the first: Turner 1892: 16–17.

138 While Turner made: Abramson 2009: 343–59; and Cullen 2006: 82–104.

140 "Turner circling": Abramson 2017: 31.

141 Its prestigious faculty: Mickens 2002: vii.

141 "The foraging bee": von Frisch 1954: 101.

142 As von Frisch painstakingly determined: von Frisch 1967.

142 "the most distinguished": Thorpe 1954: 897.

142 The astonishing conclusions: Munz 2005: 535–70.

143 After teaching at several: Munz 2016: 80–81.

144 "a petty-minded": "Political evaluation of Dr. Karl von Frisch by Wilhelm Führer," October 19, 1937, in Deichmann 1996: 42.

144 "unusually great partiality": Munz 2016: 86.

144 "Were nonhuman animals": Munz 2016: 6. Von Frisch briefly mentioned Turner's research in 1914 but remarked that he had only encountered it after having completed his own experiments (Wehner 2016: 254).

144 California's Central Valley: Mark Bittman, "Heavenly Earth," *New York Times Magazine* (October 14, 2012): MM50.

144 Like a gigantic hothouse: Jabr 2013.

145 Scientists have long warned: Kremen, Bugg, Nicola, Smith, Thorp, and Williams 2002: 41–49.

146 This tried-and-true: European Synchrotron Radiation Facility 2012.

146 Many generations prior: Fijn 56; Quezada-Euán, Nates-Parra, Maués, Imperatriz-Fonseca, and Roubik 2018: 538; and Wilson and Rhodes 2016.

146 "If the bee disappears": Einstein 2010: 479.

147 "is a tiny, rare": Hanson 2018: 89.

147 A March 2007 headline: David K. Randall, "Beekeepers Confronted by Demise of Colonies," *New York Times* (March 4, 2007): NJ7; and John Vidal, "Threat to Agriculture as Mystery Killer Wipes Out Honeybee Hives," *Guardian* (April 20, 2007).

147 "During the past season": United States Department of Agriculture 1869: 278.

148 "It remains yet a complete": Aikins 1897: 480.

148 *Varroa destructor*: Barron 2015: 45–50.

148 A controlled study: Kielmanowicz et al. 2015: e1004816.

148 Recent research has shown: Doublet, Labarussias, Miranda, Moritz, and Paxton 2015: 969–83.

148 Prior to the Second: Ujváry 1999: 29–69.

149 In addition to poisoning: Hopwood et al. 2012; and Goulson 2013: 977–87.

149 A second-order problem: Hapwood et al. 2012.

149 "the new DDT": George Monbiot, "Neonicotinoids Are the New DDT Killing the Natural World," *George Monbiot's Blog, Guardian,* August 5, 2013, https://www.theguardian.com/environment/georgemonbiot/2013/aug/05/neonicotinoids-ddt-pesticides-nature.

149 "These insects, so essential": Carson 1962: 73–74.

150 Among the most robust: Hughes 2015.

151 "Large flocks": Muir 1954: 89.

CHAPTER 8 A SIX-LEGGED MENU

152 Madagascar hissing cockroaches: Clark and Shanklin 1995.

153 "Daring guests": Great Adventure Outpost 2006.

153 The term "entomophagy": Riley 1871: 144.

153 Hailed for its potential: Evans et al. 2015: 295–96.

153 These ecological benefits: van Huis et al. 2013: xiv.

153 To take just one example: Payne, Scarborough, Rayner, and Nonaka 2016: 285–91; and Dobermann, Swift, and Field 2017: 293–308.

153 Comparing land-use needs: Martin 2014: 18–19.

153 Raising livestock: Steinfeld et al. 2006: xxi.

154 "Scientific analysis confirms": Konuma 2010: iii.

154 "Ants that burst": Charlotte Payne, "Entomophagy: How Giving Up Meat and Eating Bugs Can Help Save the Planet," *Independent* (March 21, 2018).

154 "You start with crickets": Le Vine 2017.

154 "glazed in chocolate": Witmer 2018.

154 Books with embellished titles: Martin 2014; and Ramos-Elorduy and Menzel 1998.

154 The vibrant photos: Menzel and D'Aluisio 1998.

154 Before his death: Ligaya Mishan, "Why Aren't We Eating More Insects?," *New York Times Magazine* (September 7, 2018): 84.

155 "In entering upon": Holt 1885: 5.

155 "a good red colour": Leslie 1840: 226 and 339.

155 Catharine Esther Beecher: Beecher 1846: 172 and 177.

155 "place the live grubs": Schwabe 1979: 371.

156 "First we eat": Fiona Wilson, "*The Gastronomical Me* by MFK Fisher," London *Times* (May 13, 2017).

156 Around 3400 B.C.E.: Rumold and Aldenderfer 2016: 13672–77.

156 Potatoes are part: Salaman 1949.

157 Born in 1737: Spary 2014: 167–202.

158 "frequently eat beef" and "rarely get fatigued": Shimizu 2011: 163

158 "medicinal eating," "beast markets," "mountain whale," and "winter peony": Krämer 2008: 39

158 *Casu marzu:* L'Unione Sarda 2008.

159 The Gospel of Mark: Mark 1:6, Matthew 3:4.

159 Evidence is mounting: McGrew 2014: 4–11.

159 Emboldened by the American: Cordain 2002.

160 Most humans benefit: Zhang 2018: 12–18.

160 The American evolutionary: Zuk 2013.

160 For the most part: Hardy et al. 2012: 617–26.

160 "From the time": Eaton and Nelson 1991: 281S.

161 A 2019 report: Taylor 2019.

161 Since its creation in 2013: "Little Herds HQ," http://www.littleherds.org/.

161 At least 113 countries: Jongema 2017.

161 *beondegi:* Sula 2013: 320.

161 Elsewhere in Asia: Hanboonsong 2010: 173–82.

162 *zazamushi:* Césard, Komatsu, and Iwata 2015. The primary insects grouped in the category of *zazamushi* are: stoneflies (order: Plecoptera), caddis flies (Trichoptera), dobsonflies (Megaloptera), and dragonflies and damselflies (Odonata).

162 Thick as a cigar: Dube and Dube 2010: 28–36.

163 Mopane caterpillars feed exclusively: Schiefenhövel and Blum 2007: 166; and Makhado, Potgieter, and Luus-Powell 2018: 84–90.

163 Mma Precious Ramotswe: McCall Smith 1998: 114.

164 He carried them around: Clark and Scott 2014: 78.

164 "If you try to catch": Cohen, Mata Sánchez, and Montiel-Ishino 2009: 62.

164 "I think one thing": Gomez 2018.

165 *Escamoles* resemble: Antonio Vásquez, "Meals on Wings," *Negocios ProMéxico* (December 2013–January 2014): 66–68 (statistic about price: 67).

165 "Don Bugito, the Prehispanic Snackeria": See https://www.donbugito.com/pages /about-us.

165 maguey caterpillar: Staller 2010: 39–41; Acuña, Caso, Aliphat, and Vergara 2011: 159.

165 "more as a luxury": Waugh 1916: 138–39.

166 "A fire is built": Hoffman 1878: 465.

166 "which, upon examination" and "The prejudice against": Bryant 1849: 162. In 1847, under the military rule of the United States, Bryant became the second *alcalde*—or pre-statehood mayor—of the city of San Francisco (Clark 1971: 29–43).

166 Semiannually, members: Fowler and Walter 1985: 155–65.

167 "There is no evidence": Morris 2004: 53.

167 "good advice": Clavijero 1937: 65.

167 In all likelihood: Shipek 1981: 305.

167 "Even man himself": Brues 1972: 399.

167 Factory-style chicken farming: McKenna 2017.

167 industrial-scale hog farms: Ma, Kahn, and Richt 2009: 158–66.

167 Beyond the animal cruelty: Paul Vallely, "Hugh Fearnley-Whittingstall: Crying Fowl," *Independent* (January 12, 2008).

168 "Raising insects for food": Marcel Dicke and Arnold van Huis, "The Six-Legged Meat of the Future," *Wall Street Journal* (February 19, 2011).

168 Starting in 1971: Lappé 1971.

168 The most devout vegan: United States Food and Drug Administration 2018.

168 The United Nations predicts: United Nations Department of Economic and Social Affairs 2015.

168 Half of the world's: Mintz 1985: 9; and Lampe 1995.

169 It is impossible to overstate: Hilbert et al. 2017: 1693–98.

169 As the anthropologists: Errington, Gewertz, and Fujikura 2013: 1.

169 Recently, Stockholm, Taipei: Belatchew Arkitekter, "BuzzBuilding," https://belatchew.com/en/projekt/buzzbuilding/.

169 "A lack of basic": Berggren, Jansson, and Low 2019: 132.

170 While small-sized: Lundy and Parrella 2015.

170 The mass production: Müller, Evans, Payne, and Roberts 2016: 121 36; and Dobermann, Swift, and Field 2017: 293–308.

170 the 1943 Bengal famine: Sen 1981.

170 contrasting view of the Bengal famine (footnote): Ó Gráda 2015: 38–91.

171 A team of Japanese researchers: Katayama et al. 2008: 701–05.

171 On February 20, 1947: Campbell and Garbino 2011: 470. The Kármán line, an imaginary boundary situated at an altitude of sixty-two miles above sea level, is the conventional definition of the border between Earth's atmosphere and outer space. Most international space treaties and national aerospace agencies use this line (Voosen 2018).

171 Miracle substances like algae: Belasco 1997: 608–34. Thanks to Ben Wurgaft for sharing this fascinating article with me.

171 "Some 70 percent": Du Bois 2018: 215.

172 "The aversion to insect food": Bodenheimer 1951: 10.

EPILOGUE LISTENING TO INSECTS

173 "able to tell how": Pepys 2004: 155.

173 Like a sonic fingerprint: Gross 2017.

173 "O, shrill-voiced insect": Webster 1897: 38.

174 *mushi-kiki*: Waley 2000: 173. *Ukiyo-e*—literally, "pictures of the floating world"—refers to the paintings and woodblock prints that depicted the city's pleasure districts at the time. One of these images is titled *Listening to Insects at Dokan Hill (Dōkanyama Mushikiki no Zu),* from the series *Famous Views of the Eastern Capital,* or *Tōto Meisho (Sanoki Ban),* by the artist Utagawa Hiroshige (1797–1858).

174 "The cicadas began": Revkin 1990: 1.

174 "Outdoors all afternoon": Kunitz 2005: 107–08.

175 "the stridulation of crickets": Rothenberg 2013: 2.

175 Counting a cricket's chirps: Sloane 1952: 7.

175 In Barbara Kingsolver's: Kingsolver 2012.

176 In a 2019 study, ominously titled: Sánchez-Bayo and Wyckhuys 2019: 8–27. Also: Hallmann et al. 2017. On November 27, 2018, *The New York Times Magazine* confronted this impending mass die-off with a feature story, "The Insect Apocalypse Is Here: What Does It Mean for the Rest of Life on Earth?"

176 "the hum of insects": Thoreau 1873: 48.

176 "The issue is vital": DuPuy 1925: 435.

177 "Don't accept the chauvinistic tradition": Gould 1989: 102.

177 The Anthropocene (footnote): Crutzen 2002: 23.

178 "The insect does not aim": Fabre 1911: 128. Fabre achieved far more fame in Japan than in his home country of France. Filmmaker Jessica Oreck's 2009 documentary, *Beetle Queen Conquers Tokyo*, explores the Japanese fascination with insects.

WORKS CITED

Abramson, C. I. 2017. "Charles Henry Turner Remembered." *Nature* 542: 31.

———. 2009. "A Study in Inspiration: Charles Henry Turner (1867–1923) and the Investigation of Insect Behavior." *Annual Review of Entomology* 54: 343–59.

Acuña, A. M., L. Caso, M. M. Aliphat, and C. H. Vergara. 2011. "Edible Insects as Part of the Traditional Food System of the Popoloca Town of Los Reyes Metzontla, Mexico." *Journal of Ethnobiology* 31: 150–69.

Adams, M. D. 2000. "The Genome Sequence of *Drosophila melanogaster*." *Science* 287: 2185–95.

Adams, R. 1952. "Man's Synthetic Future." *Science* 115: 157–63.

Adkins, R., and L. Adkins. 2008. *Jack Tar: The Extraordinary Lives of Ordinary Seamen in Nelson's Navy.* London: Little, Brown.

Aikin, R. C. 1897. "Bees Evaporated: A New Malady." *Gleanings in Bee Culture* 25: 479–80.

Al-Hayani, A. A., et al. 2011. "Shellac: A Non-Toxic Preservative for Human Embalming Techniques." *Journal of Animal and Veterinary Advances* 10: 1561–67.

Allen, S. 1994. *Classic Finishing Techniques.* New York: Sterling.

Almond, J. 1995. *Dictionary of Word Origins: A History of Words, Expressions, and Clichés We Use.* Secaucus, NJ: Carol.

Altman, T. A. 1998. *FDA and USDA Nutrition Labeling Guide: Decision Diagrams, Checklists, and Regulations.* Lancaster, PA: Technomic.

Ammer, C. 1989. *It's Raining Cats and Dogs . . . and Other Beastly Expressions.* New York: Paragon House.

Andrews, G. L., et al. 2008. "Dscam Guides Embryonic Axons by Netrin-Dependent and -Independent Functions." *Development* 135: 3839–48.

AnimalResearch.Info. 2019. "Nobel Prizes." http://www.animalresearch.info/en/medical-advances/nobel-prizes/.

Anonymous. 2014. "Global Demand for Polyethylene to Reach 99.6 Million Tons in 2018." *Pipeline & Gas Journal* 241. https://pgjonline.com/magazine/2014/december-2014-vol-241-no-12/features/global-demand-for-polyethylene-to-reach-996-million-tons-in-2018.

Anonymous. 2006. "The Future Imagined." *Wilson Quarterly* 30: 50–57.

Anonymous. 1922–32. *The Scriptores Historiae Augustae.* Translated by David Magie. 3 vols. New York: G. P. Putnam's Sons.

Anonymous. 1897. "Nyasa-Land." *Nature* 57: 174–75.

Anonymous. 1874. *The Happy Hour; or, Holiday Fancies and Every-day Facts for Young People.* New York: D. Appleton.

Arasaratnam, S. 1986. *Merchants, Companies, and Commerce on the Coromandel Coast, 1650–1740.* New York: Oxford University Press.

Arditti, J., J. Elliott, I. J. Kitching, and L. T. Wasserthal. 2012. "'Good Heavens What Insect Can Suck It'—Charles Darwin, *Angraecum sesquipedale* and *Xanthopan morganii praedicta.*" *Botanical Journal of the Linnaean Society* 169: 403–32.

Arditti, J., A. N. Rao, and H. Nair. 2009. "Hand-Pollination of *Vanilla:* How Many Discoverers?" In *Orchid Biology: Reviews and Perspectives, X,* edited by T. Kull, J. Arditti, and S. M. Wong. Dordrecht: Springer, 233–49.

Arrington, C. R. 1978. "The Finest Fabrics: Mormon Women and the Silk Industry in Early Utah." *Utah Historical Quarterly* 46: 376–96.

Arrizabalaga y Prado, L. de. 2010. *The Emperor Elagabalus: Fact or Fiction?* New York: Cambridge University Press.

Ashlock, P. D., and W. C. Gagné. 1983. "A Remarkable New Micropterous *Nysius* Species from the Aeolian Zone of Mauna Kea, Hawai'i Island (Hemiptera: Heteroptera: Lygaeidae)." *International Journal of Entomology* 25: 47–55.

Avila, E. 2004. "Popular Culture in the Age of White Flight: Film Noir, Disneyland, and the Cold War (Sub)Urban Imaginary." *Journal of Urban History* 31: 3–22.

Ayres, R. U. 2007. "On the Practical Limits to Substitution." *Ecological Economics* 61: 115–28.

Backus, R. L., ed. 1985. *The Riverside Counselor's Stories: Vernacular Fiction of Late Heian Japan.* Trans. R. Backus. Stanford, CA: Stanford University Press.

Balfour-Paul, J. 1998. *Indigo.* London: British Museum Press.

Bardell, D. 2004. "The Invention of the Microscope." *Bios* 75: 78–84.

Barron, A. B. 2015. "Death of the Bee Hive: Understanding the Failure of an Insect Society." *Current Opinion in Insect Science* 10: 45–50.

Barthes, R. 1972 (1957). *Mythologies.* Trans. A. Lavers. New York: Hill and Wang.

Baskes, J. 2005. "Colonial Institutions and Cross-Cultural Trade: *Repartimiento* Credit and Indigenous Production of Cochineal in Eighteenth-Century Oaxaca, Mexico." *Journal of Economic History* 65: 186–210.

———. 2000. *Indians, Merchants, and Markets: A Reinterpretation of the Repartimiento and Spanish-Indian Economic Relations in Colonial Oaxaca, 1750–1821.* Stanford, CA: Stanford University Press.

Bateson, B., ed. 1928. *William Bateson, F.R.S.: His Essays and Addresses Together with a Short Account of His Life.* Cambridge: Cambridge University Press.

Bateson, W. 1913. *Problems of Genetics.* New Haven, CT: Yale University Press.

Beattie, J. 2011. *Empire and Environmental Anxiety: Health, Science, Art and Conservation in South Asia and Australasia, 1800–1920.* New York: Palgrave Macmillan.

Beauquais, A. 1886. *Histoire économique de la soie.* Grenoble: Grands Établissements de l'Imprimerie Générale.

Beckwith, C. I. 2009. *Empires of the Silk Road: A History of Central Asia from the Bronze Age to the Present.* Princeton, NJ: Princeton University Press.

Beecher, C. 1846. *Miss Beecher's Domestic Receipt Book: Designed as a Supplement to Her Treatise on Domestic Economy.* New York: Harper.

Beezley, W. H., and M. A. Ranking, eds. 2017. *Problems in Modern Mexican History: Sources and Interpretations.* Lanham, MD: Rowman & Littlefield.

Belasco, W. 1997. "Algae Burgers for a Hungry World? The Rise and Fall of Chlorella Cuisine." *Technology and Culture* 38: 608–34.

Bell, L. S. 1999. *One Industry, Two Chinas: Silk Filatures and Peasant-Family Production in Wuxi County, 1865–1937.* Stanford, CA: Stanford University Press.

Bell, W. J., L. M. Roth, and C. A. Nalepa. 2007. *Cockroaches: Ecology, Behavior, and Natural History.* Baltimore: Johns Hopkins University Press.

Berdan, F. F., and P. R. Anawalt, eds. 1997. *The Essential Codex Mendoza.* Berkeley: University of California Press.

Berenbaum, M. 1995. *Bugs in the System: Insects and Their Impact on Human Affairs.* Reading, MA: Addison-Wesley.

Berggren, Å, A. Jansson, and M. Low. 2019. "Approaching Ecological Sustainability in the Emerging Insects-as-Food Industry." *Trends in Ecology & Evolution* 34: 132–38.

Berliner, P. F. 1994. *Thinking in Jazz: The Infinite Art of Improvisation.* Chicago: University of Chicago Press.

Bhardwaj, S. P., and R. K. Pandey. 1999–2000. "Study of Production, Trade and Policy Reform for Lac Cultivation in India." In *Encyclopedia of Agricultural Marketing,* edited by Jagdish Prasad. 12 vols. New Delhi: Mittal Publications, 227–48.

Bianco, B., R. T. Alexander, and G. Rayson. 2017. "Beekeeping Practices in Modern and Ancient Yucatán: Going from the Known to the Unknown." In *The Value of Things: Prehistoric to Contemporary Commodities in the Maya Region,* edited by Jennifer P. Mathews and Thomas H. Guderjan. Tucson: University of Arizona Press, 87–103.

Bird, W. L., Jr. 1999. *"Better Living": Advertising, Media and the New Vocabulary of Business Leadership, 1935–1955.* Evanston, IL: Northwestern University Press.

Black, F. L. 1992. "Why Did They Die?" *Science* 258: 1739–40.

Blake, W. 1977. *The Complete Poems.* Edited by A. Ostriker. Harmondsworth, UK: Penguin.

Bleichmar, D., et al., eds. 2009. *Science in the Spanish and Portuguese Empires, 1500–1800.* Stanford, CA: Stanford University Press.

Bodenheimer, F. S. 1951. *Insects as Food: A Chapter of the Ecology of Man.* The Hague: W. Junk.

Bodson, L. 1983. "The Beginnings of Entomology in Ancient Greece." *The Classical Outlook* 61: 3–6.

Boissoneault, L. 2017. "Why an Alabama Town Has a Monument Honoring the Most Destructive Pest in American History." *Smithsonian.* https://www.smithsonianmag .com/history/agricultural-pest-honored-herald-prosperity-enterprise-alabama -180963506/.

Borgström, G. 1965. *The Hungry Planet: The Modern World at the Edge of Famine.* New York: Macmillan.

Boxer, C. R. 1965. *The Dutch Seaborne Empire, 1600–1800.* New York: Alfred A. Knopf.

Bozhong, L. 1996. "From 'Husband-and-Wife Working Side-by-Side in the Fields' to 'Men Plow, Women Weave.'" *Research on Chinese Economic History* 11: 99–107.

Bradbury, R. 1962. *R Is for Rocket.* New York: Doubleday.

Bradbury, S. 1967. *The Evolution of the Microscope.* Oxford: Pergamon.

Brading, D. A. 1971. *Miners and Merchants in Bourbon Mexico, 1763–1810.* New York: Cambridge University Press.

Braudel, F. 1992. *Civilization and Capitalism: 15th–18th Century.* Translated by Siân Reynolds. 3 vols. Berkeley: University of California Press.

Bristow, I. C. 1994. "House Painting in Britain: Sources for American Paints, 1615 to 1830." In *Paint in America: The Colors of Historic Buildings,* edited by Roger W. Moss. New York: John Wiley & Sons, 42–53.

Brockway, L. H. 1979. *Science and Colonial Expansion: The Role of the British Royal Botanic Gardens.* New York: Academic Press.

Brothwell, D. R., and P. Brothwell. 1998 (1969). *Food in Antiquity: A Survey of the Diet of Early Peoples.* Baltimore: Johns Hopkins University Press.

Broven, J. 1978. *Rhythm and Blues in New Orleans.* New York: Pelican.

Brues, C. T. 1972 (1946). *Insects, Food, and Ecology.* New York: Dover.

Bryant, E. 1849. *What I Saw in California: Being the Journal of a Tour, by the Emigrant Route and South Pass of the Rocky Mountains, Across the Continent of North America, the Great Desert Basin, and Through California, in the Years 1846, 1847.* 3rd ed. New York: D. Appleton.

Bryson, B. 2010. *At Home: A Short History of Private Life.* New York: Doubleday.

Buch, K., et al. 2009. "Investigation of Various Shellac Grades: Additional Analysis for Identity." *Drug Development and Industrial Pharmacy* 35: 694–703.

Buhs, J. B. 2004. *The Fire Ant Wars: Nature, Science, and Public Policy in Twentieth-Century America.* Chicago: University of Chicago Press.

Bulnois, L. 1996. *The Silk Road.* Trans. D. Chamberlain. New York: E. P. Dutton.

Burrows, D. 2018. "America's Most Popular Breakfast Cereals (and the Stocks Behind Them)." *Kiplinger.* https://www.kiplinger.com/slideshow/investing/T052-S001-america-s-most-popular-breakfast-cereals-stocks/index.html.

BusinessWire. 2017. "Changing Fashion Trends to Boost the Global Silk Market." https://www.businesswire.com/news/home/20171206005626/en/Changing-Fashion-Trends-Boost-Global-Silk-Market.

California Academy of Sciences. 2017. "Academy Scientists Travel Down Under for Seven-Continent Exploration of Bugs in Our Homes." https://www.calacademy.org/press/releases/academy-scientists-travel-down-under-for-seven-continent-exploration-of-bugs-in-our.

California Department of Food and Agriculture. 2018. "2017 California Almond Acreage Report." https://www.nass.usda.gov/Statistics_by_State/California/Publications/Specialty_and_Other_Releases/Almond/Acreage/201804almac.pdf.

Calonius, E. 2006. *The Wanderer: The Last American Slave Ship and the Conspiracy That Set Its Sails.* New York: St. Martin's.

Campana, M. G., N. M. Robles García, and N. Tuross. 2015. "America's Red Gold: Multiple Lineages of Cultivated Cochineal in Mexico." *Ecology and Evolution* 5: 607–17.

Campbell, G. 1867. *The Reign of Law.* 2nd ed. London: Strahan.

Campbell, M. R., and A. Garbino. 2011. "History of Suborbital Spaceflight: Medical and Performance Issues." *Aviation, Space, and Environmental Medicine* 82: 469–74.

Capinera, J. L. 2008. "Harlequin Bug, *Murgantia histrionica* (Hahn) (Hemiptera: Pentatomidae)." In *Encyclopedia of Entomology*, edited by John L. Capinera. 3 vols. New York: Springer.

Carlos Rodríguez, L., and U. Pascual. 2004. "Land Clearance and Social Capital in Mountain Agro-ecosystems: The Case of Opuntia Scrubland in Ayacucho, Peru." *Ecological Economics* 49: 243–52.

Carson, R. 1998. *Lost Woods: The Discovered Writing of Rachel Carson*, edited by L. Lear. Boston: Beacon.

———. 1962. *Silent Spring*. Boston: Houghton Mifflin.

Carvalho, D. N. 1904. *Forty Centuries of Ink*. New York: Banks Law Publishing Co.

Casper, M. J., ed. 2003. *Synthetic Planet: Chemical Politics and the Hazards of Modern Life*. New York: Routledge.

Cassius, Dio. 1914–27. *Roman History*. Translated by E. Cary and H. B. Foster. 9 vols. New York: Harvard University Press.

Catton, W. R., Jr. 1980. *Overshoot: The Ecological Basis of Revolutionary Change*. Urbana: University of Illinois Press.

Cauchi, R. J. 2014. "Flying in the Face of Neurodegeneration." *Think Magazine* 8: 16–21.

Césard, N., S. Komatsu, and A. Iwata. 2015. "Processing Insect Abundance: Trading and Fishing of *Zazamushi* in Central Japan (Nagano Prefecture, Honshū Island)." *Journal of Ethnobiology and Ethnomedicine* 11. https://doi.org/10.1186/s13002-015-0066-7.

Challamel, A. 1882. *The History of Fashion in France; or, The Dress of Women from the Gallo-Roman to the Present Time*. Translated by F. C. Hoey and J. Lillie. London: Sampson, Low, Marston, Searle and Rivington.

Chamber of Commerce of the United States of America. 1921. *Our World Trade in 1920*. Washington, DC: U.S. Government Printing Office.

Chanan, M. 1995. *Repeated Takes: A Short History of Recording and Its Effects on Music*. New York: Verso.

Charter of the Forest. 1225. *The National Archives of the United Kingdom*. http://www.nationalarchives.gov.uk/education/resources/magna-carta/charter-forest-1225-westminster/.

Chatterton, E. K. 1971. *The Old East Indiamen*. London: Conway Maritime.

Chávez-Moreno, C. K., A. Tacante, and A. Casas. 2009. "The *Opuntia* (Cactaceae) and *Dactylopius* (Hemiptera: Dactylopiidae) in Mexico: A Historical Perspective of Use, Interaction and Distribution." *Biodiversity Conservation* 18: 3337–55.

Chávez Santiago, E., and H. M. Meneses Lozano. 2010. "Red Gold: Raising Cochineal in Oaxaca." *Textile Society of America Symposium Proceedings*. Paper 39. Lincoln: University of Nebraska Press.

Chetverikov, S. S. 1920. *The Fundamental Factor of Insect Evolution*. Translated by J. Kotinsky. Washington, DC: U.S. Government Printing Office.

Choi, C. Q. 2011. "Case Closed? Columbus Introduced Syphilis to Europe." *Scientific American*. https://www.scientificamerican.com/article/case-closed-columbus/.

Chouhan, T. R. 1994. *Bhopal, the Inside Story: Carbide Workers Speak Out on the World's Worst Industrial Disaster*. New York: Apex.

Christensen, T. 2012. *1619: The World in Motion*. Berkeley, CA: Counterpoint.

Clapham, M. E., and J. A. Karr. 2012. "Environmental and Biotic Controls on the Evolutionary History of Insect Body Size." *Proceedings of the National Academy of Sciences* 109: 10927–30.

Clark, D., and D. Shanklin. 1995. "ENTFACT-014: Madagascar Hissing Cockroaches (*Gromphadorhina portentosa*)." http://entomology.ca.uky.edu/ef014.

Clark, T. D. 1971. "Edwin Bryant and the Opening of the Road to California." In *Essays in Western History: In Honor of Professor T. A. Larson*, edited by R. Daniels. Laramie: University of Wyoming, 29–43.

Clark, V., and M. Scott. 2014. *Dictators' Dinners: A Bad Taste Guide to Entertaining Tyrants*. London: Gilgamesh.

Clavijero, F. J. 1937 (1789). *The History of [Lower] California*. Translated by S. E. Lake and A. A. Gray. Stanford, CA: Stanford University Press.

Cloudsley-Thompson, J. L. 1976. *Insects and History*. New York: St. Martin's.

Cobb, M. 2000. "Reading and Writing *The Book of Nature*: Jan Swammerdam (1637–1680)." *Endeavor* 24: 122–28.

Cobo, B. 1890–95 (1653). *Historia del Nuevo Mundo*, edited by M. Jiménez de la Espada. 4 vols. Sevilla: Sociedad de Bibliófilos Andaluces.

Cockerell, T.D.A. 1893. "Notes on the Cochineal Insect." *American Naturalist* 27: 1041–49.

Coe, B. 1976. *The Birth of Photography: The Story of the Formative Years, 1800–1900*. London: Ash & Grant.

Cohen, J. H., N. D. Mata Sánchez, and F. Montiel-Ishino. 2009. "*Chapulines* and Food Choices in Rural Oaxaca." *Gastronomica* 9: 61–65.

Colborn, T., D. Dumanoski, and J. P. Myers. 1996. *Our Stolen Future: Are We Threatening Our Fertility, Intelligence, and Survival? A Scientific Detective Story*. New York: E. P. Dutton.

Coll-Hurtado, A. 1998. "Oaxaca: Geografía histórica de la Grana Cochinilla." *Investigaciones geográficas boletín (Universidad Nacional Autónoma de México)* 36: 71–82.

Conlin, J. R. 2009 (1984). *The American Past: A Survey of American History*. Vol. 1, *To 1877*. 9th ed. Boston: Wadsworth.

Contreras Sánchez, A. 1987. "El palo de tinte, motivo de un conflicto entre dos naciones, 1670–1802." *Historia mexicana* 37: 49–74.

Coon, J. M., and E. A. Maynard. 1960. "Problems in Toxicology." *Federation Proceedings* (Federation of American Societies for Experimental Biology) 19: 1–52.

Cooper, B. 1989. "The House of the Future." *Grand Street* 8: 73–104.

Cordain, L. 2002. *The Paleo Diet: Lose Weight and Get Healthy by Eating the Food You Were Designed to Eat*. New York: John Wiley & Sons.

Cortés, H. 1962 (1519–26). *Five Letters of Cortés to the Emperor*. Translated by J. B. Morris. New York: W. W. Norton.

Cowan, F. 1865. *Curious Facts in the History of Insects, Including Spiders and Scorpions*. Philadelphia: J. B. Lippincott.

Crandall, E. B. 1924. *Shellac: A Story of Yesterday, Today & Tomorrow*. Chicago: James B. Day.

Crane, E. E. 1999. *The World History of Beekeeping and Honey Hunting*. New York: Routledge.

Crawford, M. S. 1859. *Life in Tuscany*. Columbus, OH: Follett, Foster.

Crosby, A. 1972. *The Columbian Exchange: The Biological and Cultural Consequences of 1492*. Westport, CT: Greenwood.

Crutzen, P. J. 2002. "Geology of Mankind." *Nature* 415: 23.

Cullen, K. E. 2006. *Biology: The People Behind the Science*. New York: Chelsea House.

Culliney, T. W. 2014. "Crop Loss Due to Arthropods." In *Integrated Pest Management*, edited by D. Pimentel and R. Peshin. Dordrecht, Netherlands: Springer, 201–25.

Dahlgren de Jordán, B. 1961. "El nocheztli o la grana de cochinilla mexicana." In *Homenaje a Pablo Martínez del Río en el vigesimoquinto aniversario de la primera edición de Los orígenes americanos*, edited by A. Caso. Mexico: Instituto Nacional de Antropología e Historia, 387–99.

Darwin, C. 1908. *Charles Darwin: His Life Told in an Autobiographical Chapter, and in a Selected Series of His Published Letters*, edited by Francis Darwin. London: John Murray.

———. 1896 (1871). *The Descent of Man, and Selection in Relation to Sex*. New York: D. Appleton.

———. 1862. "Charles Darwin to Joseph Dalton Hooker. 25 [and 26] January [1862]." *Darwin Correspondence Project*. https://www.darwinproject.ac.uk/letter/DCP-LETT-3411.xml.

———. 1862. *On the Various Contrivances by Which British and Foreign Orchids Are Fertilised by Insects, and on the Good Effects of Intercrossing*. London: John Murray.

Datta, R. K., and M. Nanavaty. 2005. *Global Silk Industry: A Complete Sourcebook*. Boca Raton, FL: Universal.

Datta, S. 2010. "Cockroaches: Defying the Passage of Time." *Science Reporter* 47: 55.

Dave, K. N. 1950. *Lac and the Lac-insect in the Athara-Veda*. Nagpur: International Academy of Indian Culture.

Davis, D. L. 1998. *The Secret History of the War on Cancer*. New York: Basic Books.

Davis, F. H. 1912. *Myths and Legends of Japan*. London: Ballantyne.

Davis, F. R. 2008. "Unraveling the Complexities of Joint Toxicity of Multiple Chemicals at the Tox Lab and the FDA." *Environmental History* 13: 674–83.

Davis, M. 2001. *Late Victorian Holocausts: El Niño Famines and the Making of the Third World*. New York: Verso.

Davis, N. Z. 2011. "Judges, Masters, Diviners: Slaves' Experience of Criminal Justice in Colonial Suriname." *Law and History Review* 29: 925–84.

———. 1995. *Women on the Margins: Three Seventeenth-Century Lives*. Cambridge, MA: Harvard University Press.

Day, T. 2000. *A Century of Recorded Music: Listening to Musical History*. New Haven, CT: Yale University Press.

Deichmann, U. 1996. *Biologists Under Hitler*. Translated by Thomas Dunlap. Cambridge, MA: Harvard University Press.

DeLong, D. M. 1960. "Man in a World of Insects." *Ohio Journal of Science* 60: 193–206.

Denning, M. 2015. *Noise Uprising: The Audiopolitics of a World Musical Revolution.* New York: Verso.

Deshpande, S. S. 2002. *Handbook of Food Toxicology.* New York: Marcel Dekker.

Dickinson, E. 2003. *The Collected Poems of Emily Dickinson,* edited by R. Wetzsteon. New York: Barnes & Noble Classics.

———. 1924. *The Complete Poems of Emily Dickinson.* Boston: Little, Brown.

Dillard, A. 1974. *Pilgrim at Tinker Creek.* New York: Harper's Magazine Press.

Diouf, S. A. 2007. *Dreams of Africa in Alabama: The Slave Ship* Clotilda *and the Story of the Last Africans Brought to America.* New York: Oxford University Press.

Dobermann, D., J. A. Swift, and L. M. Field. 2017. "Opportunities and Hurdles of Edible Insects for Food and Feed." *Nutrition Bulletin* 42: 293–308.

Dodds, B. 1992. *The Baby Dodds Story.* Rev. ed. Baton Rouge: Louisiana State University Press.

Donkin, R. A. 1977. *Spanish Red: An Ethnogeographical Study of Cochineal and the Opuntia Cactus.* Philadelphia: American Philosophical Society.

Donne, J. 1971. *John Donne: The Complete English Poems,* edited by A. J. Smith. New York: Penguin.

Doublet, V., M. Labarussias, J. R. Miranda, R.F.A. Moritz, and R. J. Paxton. 2015. "Bees Under Stress: Sublethal Doses of a Neonicotinoid Pesticide and Pathogens Interact to Elevate Honey Bee Mortality Across the Life Cycle." *Environmental Microbiology* 17: 969–83.

Dow, G. F. 1927. *The Arts and Crafts of New England 1704–1775.* Topsfield, MA: Wayside.

Dowd, T. 2006. "From 78s to MP3s: The Embedded Impact of Technology in the Market for Prerecorded Music." In *The Business of Culture: Strategic Perspectives on Entertainment and Media,* edited by J. Lampel, J. Shamsie, and T. K. Lant. New York: Routledge, 205–24.

Downham, A., and P. Collins. 2000. "Colouring Our Foods in the Last and Next Millennium." *International Journal of Food Science and Technology* 35: 5–22.

Drayton, R. H. 2000. *Nature's Government: Science, Imperial Britain, and the "Improvement" of the World.* New Haven, CT: Yale University Press.

Dube, S., and C. Dube. 2010. "Towards Improved Utilization of Macimbi *Imbrasia belina* Linnaeus, 1758 as Food and Financial Resource for People in the Gwanda District of Zimbabwe." *Zimbabwe Journal of Science and Technology* 5: 28–36.

Du Bois, C. M. 2018. *The Story of Soy.* London: Reaktion.

Dudley, R. 2000. *The Biomechanics of Insect Flight: Form, Function, Evolution.* Princeton, NJ: Princeton University Press.

DuPuy, W. A. 1925. "The Insects Are Winning: A Report on the Thousand-Year War." *Harper's Magazine,* 435–40.

Durr, K. D. 2006. "The 'New Industrial Philosophy': U.S. Corporate Recycling in World War II." *Progress in Industrial Ecology* 3: 361–78.

Durrant, S. W. 1995. *The Cloudy Mirror: Tension and Conflict in the Writings of Sima Qian.* Albany: State University of New York Press.

Earth, B., et al. 2008. "Intensification Regimes in Village-Based Silk Production, Northeast Thailand: Boosts (and Challenges) to Women's Authority." In *Gender and Natural Resource Management: Livelihoods, Mobility and Interventions,* edited by B. P. Resurreccion and R. Elmhirst. Sterling, VA: Earthscan, 43–66.

Eaton, S. B., and D. A. Nelson. 1991. "Calcium in Evolutionary Perspective." *American Journal of Clinical Nutrition* 54: 281S–87S.

Edwardes, Charles. 1888. *Rides and Studies in the Canary Islands.* London: T. Fisher Unwin.

Eiland, E. 2003 (2000). *Oriental Rugs Today: A Guide to the Best New Carpets from the East.* Albany, CA: Berkeley Hills Books.

Einstein, A. 2010. *The Ultimate Quotable Einstein,* edited by A. Calaprice. Princeton, NJ: Princeton University Press.

Eisner, T. 2005. *For the Love of Insects.* Cambridge, MA: Belknap.

Eisner, T., S. Nowicki, M. Goetz, and J. Meinwald. 1980. "Red Cochineal Dye (Carminic Acid): Its Role in Nature." *Science* 208: 1039–42.

Ejsmont, R. K., and B. A. Hassan. 2014. "The Little Fly That Could: Wizardry and Artistry of *Drosophila* Genomics." *Genes* 5: 385–414.

Elisseeff, V. 1998. "Approaches Old and New to the Silk Roads." In *Silk Roads: Highways of Culture and Commerce,* edited by V. Elisseeff. New York: Berghahn.

Endersby, J. 2008. *Imperial Nature: Joseph Hooker and the Practices of Victorian Science.* Chicago: University of Chicago Press.

Engel, M. S. 2018. *Innumerable Insects: The Story of the Most Diverse and Myriad Animals on Earth.* New York: Sterling.

Enright, M. J. 1996. *Lady with a Mead Cup: Ritual, Prophecy, and Lordship in the European Warband from La Tène to the Viking Age.* Portland, OR: Four Courts.

Errington, F., D. Gewertz, and T. Fujikura. 2013. *The Noodle Narratives: The Global Rise of an Industrial Food into the Twenty-First Century.* Berkeley: University of California Press.

Etheridge, K. 2011. "Maria Sibylla Merian: The First Ecologist?" In *Women and Science, 17th Century to Present: Pioneers, Activists and Protagonists,* edited by Donna Spalding Adreolle and Veronique Molinari. Newcastle, UK: Cambridge Scholars, 35–54.

Eugenides, J. 2002. *Middlesex.* New York: Farrar, Straus and Giroux.

European Synchrotron Radiation Facility. 2012. "First Ever Record of Insect Pollination from 100 Million Years Ago." *ScienceDaily.* https://www.sciencedaily.com/releases/2012/05/120514153113.htm.

Evans, J., et al. 2015. "'Entomophagy': An Evolving Terminology in Need of Review." *Journal of Insects as Food and Feed* 1: 293–305.

Fabre, J.-H. 1911. *The Life and Love of the Insect.* Translated by Alexander Teixeira de Mattos. New York: Macmillan.

———. 1879–1907. *Souvenirs entomologiques,* edited by Yves Delange. 10 vols. Paris: Robert Laffont.

Fardisi, M., A. D. Gondhalekar, A. R. Ashbrook, and M. E. Scharf. 2019. "Rapid Evolutionary Reponses to Insecticide Resistance Management Interventions by the German Cockroach (*Blattella germanica* L.)." *Science Reports* 9. https://doi.org/10.1038/s41598-019-44296-y.

Fatah-Black, K. J. 2013. "Suriname and the Atlantic World, 1650–1800." PhD thesis, Leiden University.

Fazlýodlu, A., and O. Aslanapa. 2006. *The Last Loop of the Knot: Ottoman Court Carpets.* Istanbul: TBMM.

Feig, V. R., H. Tran, and Z. Bao. 2018. "Biodegradable Polymeric Materials in Degradable Electronic Devices." *ACS Central Science* 4: 337–48.

Fenichell, S. 1996. *Plastic: The Making of a Synthetic Century.* New York: Harper Business.

Ferrer, Eulalio. 2007 (1999). *Los lenguajes del color.* Mexico: Fondo de Cultura Económica.

Fijn, N. 2014. "Sugarbag Dreaming: The Significance of Bees to Yolngu in Arnhem Land, Australia." *Humanimalia* 6: 41–61.

Fijn, N., and M. Baynes-Rock. 2018. "A Social Ecology of Stingless Bees." *Human Ecology* 46: 207–16.

Finlay, M. R. 2009. *Growing American Rubber: Strategic Plants and the Politics of National Security.* New Brunswick, NJ: Rutgers University Press.

Firn, R. 2010. *Nature's Chemicals: The Natural Products That Shaped Our World.* New York: Oxford University Press.

Fisher, R. A. 1936. "Has Mendel's Work Been Rediscovered?" *Annals of Science* 1: 115–26.

Fisher, R. A., and G. R. De Beer. 1947. "Thomas Hunt Morgan, 1866–1945." *Obituary Notices of Fellows of the Royal Society* 5: 451–66.

Flick, D. 2006. *Africa: Continent of Economic Opportunity.* Johannesburg, South Africa: STE Publishers.

Fong, G. S. 2004. "Female Hands: Embroidery as a Knowledge Field in Women's Everyday Life in Late Imperial and Early Republican China." *Late Imperial China* 25: 1–58.

Food and Agriculture Organization of the United Nations. 2016. "Pollinators Vital to Our Food Supply Under Threat." http://www.fao.org/news/story/en/item/384726/icode/.

Foster, J. B. 2005. "The Vulnerable Planet." In *Environmental Sociology: From Analysis to Action,* edited by L. King and D. McCarthy. Lanham, MD: Rowman & Littlefield, 3–15.

Fowler, C. S., and N. P. Walter. 1985. "Harvesting Pandora Moth Larvae with the Owens Valley Paiute." *Journal of California and Great Basin Anthropology* 7: 155–65.

Fox-Davies, A. C. 2007 (1969). *A Complete Guide to Heraldry.* New York: Bonanza.

Franceschini, N., J. M. Pichon, and C. Blanes. 1992. "From Insect Vision to Robot Vision." *Philosophical Transactions of the Royal Society of London, Series B: Biological Sciences* 337: 283–94.

Francis, C., ed. 2003. *There Is Nothing Like a Thane! The Lighter Side of "Macbeth."* New York: Thomas Dunne.

Frank, E. T., et al. 2017. "Saving the Injured: Rescue Behavior in the Termite-Hunting Ant *Megaponera analis.*" *Science Advances* 3. DOI: 10.1126/sciadv.1602187.

Frankopan, P. 2017. *The Silk Roads: A New History of the World.* New York: Vintage.

Frickman Young, C. E. 2003. "Socioeconomic Causes of Deforestation in the Atlantic

Forest of Brazil." In *The Atlantic Forest of South America: Biodiversity Status, Threats, and Outlook,* edited by C. Galindo Leal and I. de Gusmão Câmara. Washington, DC: Island Press.

Friedewald, B. 2015. *A Butterfly Journey: Maria Sibylla Merian, Artist and Scientist.* Translated by S. von Pohl. New York: Prestel.

Frith, S. 2004. *Popular Music: Critical Concepts in Media and Cultural Studies.* New York: Routledge.

Gajanan, D. 2009. "Ahimsa Peace Silk—An Innovation in Silk Manufacturing." *Man-made Textiles in India* 52: 421–24.

Galeano, E. 1987. *The Memory of Fire.* Vol. 2, *Faces and Masks.* 3 vols. New York: Pantheon.

Geetha, G. S., and R. Indira. 2011. "Silkworm Rearing by Rural Women in Karnataka: A Path to Empowerment." *Indian Journal of Gender Studies* 18: 89–102.

George, T. S. 2001. *Minamata: Pollution and the Struggle for Democracy in Postwar Japan.* Cambridge, MA: Harvard University Press.

Geyer, R., J. R. Jambeck, and K. L. Law. 2017. "Production, Use, and Fate of All Plastics Ever Made." *Science Advances* 3. https://doi.org/10.1186/s13002-015-0066-7.

Ghosh, M. K. 2015. "Lac Industry in India: A Momentary View." *Open Eyes: Indian Journal of Social Science, Literature, Commerce, & Allied Areas* 12: 105.

Gibbs, L. M. (as told to M. Levine). 1982. *Love Canal: My Story.* New York: Grove.

Giesen, J. C. 2011. *Boll Weevil Blues: Cotton, Myth, and Power in the American South.* Chicago: University of Chicago Press.

Ginsberg, M. 2007. "Donatella & Allegra." *Harper's Bazaar.* https://www.harpersbazaar .com/fashion/designers/a82/donatella-allegra-versace-0307/.

Glickman, L. B. 2005. " 'Make Lisle the Style': The Politics of Fashion in the Japanese Silk Boycott, 1937–1940." *Journal of Social History* 38: 573–608.

Global Market Insights, Inc. 2016. "Edible Insects Market Size Worth $522mn by 2023." https://www.gminsights.com/pressrelease/edible-insects-market.

Goff, M. L. 2000. *A Fly for the Prosecution: How Insect Evidence Helps Solve Crimes.* Cambridge, MA: Harvard University Press.

Goldsmith, J. L., and T. Wu. 2006. *Who Controls the Internet? Illusions of a Borderless World.* New York: Oxford University Press.

Gomez, E. 2018. "A Side of Grasshoppers." *ESPN.com.* http://www.espn.com/espn/feature /story/ /id/22946221/at-seattle-mariners-games-grasshoppers-favorite-snack.

Gómez de Cervantes, G. 1944 (1599). *Vida económica y social de Nueva España al finalizar el siglo XVI,* edited by A. María Carreño. Mexico: Antigua Librería Robredo.

González Lemus, N. 2001. "La explotación de la cochinilla en las Canarias del siglo XIX." *Arquipélago-História* 5: 175–92.

Goslinga, C. 1979. *A Short History of the Netherlands Antilles and Surinam.* The Hague: M. Nijhoff.

Gould, S. J. 1989. *Wonderful Life: The Burgess Shale and the Nature of History.* New York: W. W. Norton.

Goulson, D. 2013. "Review: An Overview of the Environmental Risks Posed by Neonicotinoid Insecticides." *Journal of Applied Ecology* 50: 977–87.

Grabowski, C. 2017. *Maria Sibylla Merian zwischen Malerei und Naturforschung* (Maria Sibylla Merian Between Painting and Natural Science). Berlin: Reimer Verlag.

Granata, C. L. 2002. "The Battle for the Vinyl Frontier." In *45 RPM: A Visual History of the Seven-Inch Record*, edited by Spencer Drate. Princeton, NJ: Princeton Architectural Press, 6–11.

Grand View Research. 2018. "Carmine Market Size, Share & Trends Analysis Report By Application (Beverages, Bakery & Confectionary, Dairy & Frozen Products, Meat, Oil & Fat), By Region, And Segment Forcasts, 2018–2025." https://www.grandview research.com/industry-analysis/carmine-market.

Great Adventure Outpost. 2006. "Six Flags' 'Fright Fest' Halloween Celebration Offers Guests Both Tricks AND Treats." http://www.gadvoutpost.com/index .php?topic=911.0.

Greenfield, A. B. 2005. *A Perfect Red: Empire, Espionage, and the Quest for the Color of Desire*. New York: HarperCollins.

Greenspan, R. J. 1997. *Fly Pushing: The Theory and Practice of* Drosophila *Genetics*. Plainview, NY: Cold Spring Harbor Laboratory Press.

Grimaldi, D., and M. S. Engel. 2005. *Evolution of the Insects*. New York: Cambridge University Press.

Gross, D. A. 2017. "Why We Need to Start Listening to Insects." *Smithsonian Magazine*. https://www.smithsonianmag.com/science-nature/why-we-need-start-listening -insects-180963014/.

Grove, R. H. 1995. *Green Imperialism: Colonial Expansion, Tropical Island Edens and the Origins of Environmentalism, 1600–1860*. New York: Cambridge University Press.

Grupen, C., and M. Rodgers. 2016. *Radioactivity and Radiation: What They Are, What They Do, and How to Harness Them*. Cham, Switzerland: Springer.

Gullan, P. J., and P. S. Cranston. 2014 (2000). *The Insects: An Outline of Entomology*. Hoboken, NJ: John Wiley & Sons.

Gustafsson, Å. 1969. "The Life of Gregor Johann Mendel—Tragic or Not?" *Hereditas* 62: 239–58.

Hahn, O., W. Malzer, B. Kanngiesser, and B. Beckhoff. 2004. "Characterization of Iron-Gall Inks in Historical Manuscripts and Music Compositions Using X-Ray Fluorescence Spectrometry." *X-Ray Spectrometry* 33: 234–39.

Hakluyt, R. 1903. *The Principal Navigations, Voyages, Traffiques & Discoveries of the English Nation*. 12 vols. New York: Macmillan.

Hallman, C. A., et al. 2017. "More Than 75 Percent Decline over 27 Years in Total Flying Insect Biomass in Protected Areas." *PLOS One* 12. https://doi.org/10.1371/journal .pone.0185809.

Halloran, A., R. Flore, P. Vantomme, and N. Roos, eds. 2018. *Edible Insects in Sustainable Food Systems*. Cham, Switzerland: Springer.

Hamilton, F. 1807. *A Journey from Madras Through the Countries of Mysore, Canara, and Malabar*. 3 vols. London: T. Cadell and W. Davies.

Hammers, R. 1998. "The Fabrication of Good Government: Images of Silk Production in Southern Song (1127–1279) and Yuan (1279–1368) China." *Textile Society of America Symposium Proceedings* 171: 195–203.

Hamowy, R. 2007. *Government and Public Health in America.* Northampton, MA: Edward Elgar.

Hanboonsong, Y. 2010. "Edible Insects and Associated Food Habits in Thailand." In *Forest Insects as Food: Humans Bite Back,* edited by P. B. Durst, et al. Bangkok: Food and Agriculture Organization of the United Nations, Regional Office for Asia and the Pacific, 173–82.

Hanson, T. 2018. *Buzz: The Nature and Necessity of Bees.* New York: Basic Books.

Hardy, K., et al. 2012. "Neandertal Medics? Evidence for Food, Cooking, and Medicinal Plants Entrapped in Dental Calculus." *Naturwissenschaften* 99: 617–26.

Hargett, J. M. 2006. *Stairway to Heaven: A Journey to the Summit of Mount Emei.* Albany: State University of New York Press.

Harpp, K. 2002. "How Do Volcanoes Affect World Climate?" *Scientific American.* https://www.scientificamerican.com/article/how-do-volcanoes-affect-w/.

Harris, S. H. 1994. *Factories of Death: Japanese Biological Warfare, 1932–45, and the American Cover-up.* New York: Routledge.

Harvey, R., and M. R. Mahard. 2014. *The Preservation Management Handbook: A 21st-Century Guide for Libraries, Archives, and Museums.* Lanham, MD: Rowman & Littlefield.

Hatch, C. E., Jr. 1957. "Mulberry Trees and Silkworms: Sericulture in Early Virginia." *Virginia Magazine of History and Biography* 65: 3–61.

Hearn, L. 1899. *In Ghostly Japan.* Boston: Little, Brown.

Henderson, H. G. 1958. *An Introduction to Haiku: An Anthology of Poems and Poets from Basho to Shiki.* New York: Doubleday.

Henig, R. M. 2000. *The Monk in the Garden: The Lost and Found Genius of Gregor Mendel, the Father of Genetics.* Boston: Houghton Mifflin.

Herber, L. [M. Bookchin]. 1962. *Our Synthetic Environment.* New York: Alfred A. Knopf.

Hermes, M. E. 1996. *Enough for One Lifetime: Wallace Carothers, Inventor of Nylon.* Washington, DC: Chemical Heritage Foundation.

Hertz, G. B. 1909. "The English Silk Industry in the Eighteenth Century." *English Historical Review* 24: 710–27.

Hesse-Honegger, C. 2001. *Heteroptera: The Beautiful and the Other; or, Images of a Mutating World.* Translated by C. Luisi. New York: Scalo.

Hilbert, L., et al. 2017. "Evidence for Mid-Holocene Rice Domestication in the Americas." *Nature Ecology & Evolution* 11 1693 98.

Hill, J. E. 2009. *Through the Jade Gate to Rome: A Study of the Silk Routes During the Later Han Dynasty 1st to 2nd Centuries CE.* Charleston, SC: BookSurge.

Hoang-Dao, B.T., et al. 2009. "Clinical Efficiency of a Natural Resin Fluoride Varnish (Shellac F) in Reducing Dentin Hypersensitivity." *Journal of Oral Rehabilitation* 36: 124–31.

Hoare, B. 2009. *Animal Migration: Remarkable Journeys in the Wild.* Berkeley: University of California Press.

Hofenk–De Graaff, J. H. 1983. "The Chemistry of Red Dyestuffs in Medieval and Early Modern Europe." In *Cloth and Clothing in Medieval Europe,* edited by N. B. Harte and K. G. Ponting. London: Heinemann, 71–79.

Hoffman, W. J. 1878. "Miscellaneous Ethnographic Observations on Indians Inhabiting Nevada, California, and Arizona." In *Tenth Annual Report of the United States Geological and Geographical Survey of the Territories—Being a Report of Progress of the Exploration for the Year 1876*. Washington, DC: Government Printing Office, 465–66.

Hofstra, W. R. 2004. *The Planting of New Virginia: Settlement and Landscape in the Shenandoah Valley*. Baltimore: Johns Hopkins University Press.

Holden, C. 2006. "Report Warns of Looming Pollination Crisis in North America." *Science* 314: 397.

Holland, C., F. Vollrath, A. J. Ryan, and O. O. Mykhaylvk. 2012. "Silk and Synthetic Polymers: Reconciling 100 Degrees of Separation." *Advanced Materials* 24: 105–09.

Holt, V. M. 1885. *Why Not Eat Insects?* London: Field & Tuer, Leadenhall Press.

Hoogbergen, W.S.M. 2008. *Out of Slavery: A Surinamese Roots History*. Berlin: Lit Verlag.

Hopkirk, P. 1980. *Foreign Devils on the Silk Road: The Search for the Lost Cities and Treasures of Central Asia*. London: Murray.

Hopwood, J., et al. 2012. *Are Neonicotinoids Killing Bees? A Review of Research into the Effects of Neonicotinoid Insecticides on Bees, with Recommendations for Action*. Xerces Society for Invertebrate Conservation. http://cues.cfans.umn.edu/old/pollinators/pdf-pesticides/Are-Neonicotinoids-Killing-Bees_Xerces-Society.pdf.

Horn, T. 2005. *Bees in America: How the Honey Bee Shaped a Nation*. Lexington: University Press of Kentucky.

Hornborg, A. 2001. *The Power of the Machine: Global Inequalities of Economy, Technology, and Environment*. Walnut Creek, CA: AltaMira.

Houston, K. 2016. *The Book: A Cover-to-Cover Exploration of the Most Powerful Object of Our Time*. New York: W. W. Norton.

Hoyt, E., and T. Schultz, eds. 1999. *Insect Lives: Stories of Mystery and Romance from a Hidden World*. Edinburgh: Mainstream.

Hublin, J.-J., et al. 2017. "New Fossils from Jebel Irhoud, Morocco and the Pan-African Origin of *Homo sapiens*." *Nature* 546: 289–92.

Hudson, G. F. 1931. *Europe and China: A Survey of Their Relations from the Earliest Times to 1800*. London: Edward Arnold.

Hughes, H. J. 2015. "Reflections on 50 Years of Engagement with the Natural World: Interview with Robert Michael Pyle." *Terrain.org*. https://www.terrain.org/2015/interviews/robert-michael-pyle/.

Hughes, T. P. 1989. *American Genesis: A Century of Invention and Technological Enthusiasm, 1870–1970*. New York: Penguin Books.

Hugo, V. 1915. *Les Misérables*. Translated by I. F. Hapgood. New York: Thomas Y. Crowell.

Hume, D. 1956 (1757). *The Natural History of Religion*, edited by H. E. Root. London: A. and C. Black Ltd.

Huq, A., et al. 2010. "Simple Sari Cloth Filtration of Water Is Sustainable and Continues to Protect Villagers from Cholera in Matlab, Bangladesh." *mBio* 1: 1–5.

Hutchinson, G. E. 1977. In "The Influence of the New World on the Study of Natural History." *Changing Scenes in the Natural Sciences, 1776–1976*, edited by Clyde E. Goulden. Philadelphia: Academy of Natural Sciences, 13–34.

————. 1959. "Homage to Santa Rosalia; or, Why Are There So Many Kinds of Animals?" *American Naturalist* 93: 145–59.

Ibn Khallikan. 1843–71. *Ibn Khallikan's Biographical Dictionary.* Translated by B.M.G. de Slane. 4 vols. Paris: Oriental Translation Fund of Great Britain and Northern Ireland.

Illica, L., G. Giacosa, R. H. Elkin, and G. Puccini. 1906. *Madam Butterfly: A Japanese Tragedy.* Founded on the book by John L. Long and the drama by David Belasco. 2nd ed. London: G. Ricordi.

Imbarex Natural Colors & Ingredients. 2019. "Which Food Products Contain Carmine?" https://www.imbarex.com/wich-food-products-contain-carmine/.

India Brand Equity Foundation. 2018. "Shellac And Forest Products Industry & Exports In India." https://www.ibef.org/exports/shellac-forest-products-industry-india.aspx.

Inward, D., G. Beccaloni, and P. Eggleton. 2007. "Death of an Order: A Comprehensive Molecular Phylogenetic Study Confirms That Termites Are Eusocial Cockroaches." *Biology Letters* 3: 331–35.

Irimia-Vladu, M., et al. 2013. "Natural resin shellac as a substrate and a dielectric layer for organic field-effect transistors." *Green Chemistry* 15: 1473–76.

Islam, I., and M. Hossain. 2006. *Globalisation and the Asia-Pacific: Contested Perspectives and Diverse Experience.* Northampton, MA: Edward Elgar.

Jabr, F. 2013. "The Mind-Boggling Math of Migratory Beekeeping." https://www.scientific american.com/article/migratory-beekeeping-mind-boggling-math/.

Jaffe, B. 1976 (1948). *Crucibles: The Story of Chemistry from Ancient Alchemy to Nuclear Fission.* 4th ed. New York: Dover.

James, C.L.R. 1938. *The Black Jacobins: Toussaint L'Ouverture and the San Domingo Revolution.* London: Secker & Warburg.

Janerich, D. T., et al. 1981. "Cancer Incidence in the Love Canal Area." *Science* 212: 1404–07.

Jaubert, S., A. Mereau, C. Antoniewski, and D. Tagu. 2007. "MicroRNAs in *Drosophila*: The Magic Wand to Enter the Chamber of Secrets?" *Biochimie* 89: 1211–20.

Jenkins, M. 2011. *ContamiNation: My Quest to Survive in a Toxic World.* New York: Avery.

Jones, J. B. 2006. *The Songs that Fought the War: Popular Music and the Home Front, 1939–1945.* Waltham, MA: Brandeis University Press.

Jongema, Y. 2017. "List of Edible Insects of the World." Department of Entomology, Wangeningen University and Research. https://www.wur.nl/en/Research-Results /Chair-groups/Plant-Sciences/Laboratory-of-Entomology/Edible-insects/World wide-species-list.htm.

Joyce, J. 1939. *Finnegans Wake.* London: Faber & Faber.

Kadavy, D. R., et al. 1999. "Microbiology of the Oil Fly, *Helaeomyia petrolei.*" *Applied and Environmental Microbiology* 65: 1477–82.

Kafka, F. 2014 (1915). *The Metamorphosis.* Translated by Susan Bernofsky. New York: W. W. Norton.

Kahane, R., et al. 2008. "Bourbon Vanilla: Natural Flavour with a Future." *Chronica Horticulturae* 48: 23–29.

Kameswari, V.L.V. 2004. "Communication Network in Forest Management: Privileging Men's Voices over Women's Knowledge." *Gender Technology and Development* 8: 167–83.

Kanfer, S. 2000. *Groucho: The Life and Times of Julius Henry Marx.* New York: Alfred A. Knopf.

Katayama, N., et al. 2008. "Entomophagy: A Key to Space Agriculture." *Advances in Space Research* 41: 701–05.

Katz, M. 2010. *Capturing Sound: How Technology Has Changed Music.* Berkeley: University of California Press.

Kaufman, L., and A. Kaufman. 2003. *A Fiddler's Tale: How Hollywood and Vivaldi Discovered Me.* Madison: University of Wisconsin Press.

Keenan, K. 1983. "Lilian Vaughan Morgan (1870–1952): Her Life and Work." *American Zoologist* 23: 867–76.

Keene, J. H. 1891. *Fly-fishing and Fly-making for Trout, Bass, Salmon, Etc.* New York: Forest and Stream.

Keller, A. 2007. "*Drosophila melanogaster*'s History as a Human Commensal." *Current Biology* 17: R77–R81.

Kerr, J., and J. Banks. 1781. "Natural History of the Insect Which Produces the Gum Lacca. By Mr. James Kerr, of Patna; Communicated by Sir Joseph Banks, P.R.S." *Philosophical Transactions of the Royal Society of London* 71: 374–82.

Kiauta, M. 1986. "Dragonfly in Haiku." *Odonatologica* 15: 91–96.

Kielmanowicz, M. G., et al. 2015. "Prospective Large-Scale Field Study Generates Predictive Model Identifying Major Contributors to Colony Losses." *PLoS Pathogens* 11. http://dx.doi.org/10.1371/journal.ppat.1004816.

Killeffer, D. H. 1943. "Promise and Problems of Peace: Chemical Industry's Postwar Role." *Industrial and Engineering Chemistry* 35: 1140–45.

Kingsolver, B. 2012. *Flight Behavior: A Novel.* New York: Harper.

Kinkela, D. 2011. *DDT and the American Century: Global Health, Environmental Politics, and the Pesticide That Changed the World.* Chapel Hill: University of North Carolina Press.

Kinne-Saffran, E., and R.K.H. Kinne. 1999. "Vitalism and Synthesis of Urea: From Friedrich Wöhler to H. A. Krebs." *American Journal of Nephrology* 19: 290–94.

Kirby, J., M. Spring, and C. Higgitt. 2007. "The Technology of Eighteenth- and Nineteenth-Century Red Lake Pigments." *National Gallery Technical Bulletin* 28: 69–95.

Klein, A.-M., et al. 2007. "Importance of Pollinators in Changing Landscapes for World Crops." *Proceedings of the Royal Society B* 274: 303–13.

Klose, N. 1963. "Sericulture in the United States." *Agricultural History* 27: 225–34.

Knaut, A. L. 1997. "Yellow Fever and the Late Colonial Public Health Response in the Port of Veracruz." *Hispanic American Historical Review* 77: 619–44.

Kohler, R. 1994. *Lords of the Fly: Drosophila Genetics and the Experimental Life.* Chicago: University of Chicago Press.

Kolbert, E. 2014. *The Sixth Extinction: An Unnatural History.* New York: Henry Holt.

Konuma, H. "Foreward." 2010. In *Forest Insects as Food: Humans Bite Back,* edited by P. B. Durst, et al. Bangkok: Food and Agriculture Organization of the United Nations, Regional Office for Asia and the Pacific, iii.

Krainik, C., M. Krainick, and C. Walvoord. 1988. *Union Cases: A Collector's Guide to the Art of America's First Plastics.* Grantsburg, WI: Centennial Photo Service.

Krämer, H. M. 2008. "'Not Befitting Our Divine Country': Eating Meat in Japanese Discourses of Self and Other from the Seventeenth Century to the Present." *Food and Foodways* 16: 33–62.

Kremen, C., R. L. Bugg, N. Nicola, S. A. Smith, R. W. Thorp, and N. M. Williams. 2002. "Native Bees, Native Plants and Crop Pollination in California." *Fremontia* 30: 41–49.

Kritsky, G. 2015. *The Tears of Re: Beekeeping in Ancient Egypt.* New York: Oxford University Press.

———. 1991. "Darwin's Madagascan Hawk Moth Prediction." *American Entomologist* 37: 206–10.

Kuhn, D. 1984. "Tracing a Chinese Legend: In Search of the Identity of the 'First Sericulturalist.'" *T'oung Pao* 70: 213–45.

Kunitz, S. 2005. *The Wild Braid: A Poet Reflects on a Century in the Garden.* New York: W. W. Norton.

Kyo, C. 2012. *The Search for the Beautiful Woman: A Cultural History of Japanese and Chinese Beauty.* Translated by K. I. Selden. Lanham, MD: Rowman & Littlefield.

La Point, R. 2012. *Oshkosh: Preserving the Past.* Indianapolis, IN: Dog Ear Publishing.

Laërtius, D. 1853. *Lives and Opinions of Eminent Philosophers.* Translated by C. D. Yonge. London: H. G. Bohn.

Lampe, K. 1995. "Rice Research: Food for 4 Billion People." *GeoJournal* 35: 253–61.

Langmuir, A. C. 1915. "Shellac." In *Industrial Chemistry,* edited by A. Rogers. 2nd ed. New York: D. Van Nostrand Co.

Lappé, F. M. 1971. *Diet for a Small Planet.* New York: Ballantine.

Latour, B. 1993. *We Have Never Been Modern.* Translated by C. Porter. Cambridge, MA: Harvard University Press.

Lawry, J. V. 2006. *The Incredible Shrinking Bee: Insects as Models for Microelectromechanical Devices.* London: Imperial College Press.

Leal-Egaña, A., and T. Scheibel. 2010. "Silk-Based Materials for Biomedical Applications." *Biotechnology and Applied Biochemistry* 55: 155–67.

Lear, L. 1989. *Rachel Carson: Witness for Nature.* New York: Henry Holt.

Le Corbusier [C.-E. Jeanneret]. 1927 (1923). *Towards a New Architecture.* Translated by F. Etchells. New York: Dover Publications.

Le Coz, C.-J., et al. 2002. "Allergic Contact Dermatitis from Shellac in Mascara." *Contact Dermatitis* 46: 149–52.

Lee, R. 1951. "American Cochineal in European Commerce, 1526–1625." *Journal of Modern History* 23: 205–24.

Leggett, W. F. 1944. *Ancient and Medieval Dyes.* Brooklyn: Chemical Publishing Co.

Lehane, M. 2005. *The Biology of Blood-Sucking in Insects.* 2nd ed. New York: Cambridge University Press.

Leong, M., et al. 2017. "The Habitats Humans Provide: Factors Affecting the Diversity and Composition of Arthropods in Houses." *Nature: Scientific Reports* 7: 1–11.

Leslie, E. 1840. *Directions for Cookery, in Its Various Branches.* Philadelphia: E. L. Carey & Hart.

Le Vine, L. 2017. "Watch Angelina Jolie and Her Children Cook and Eat Bugs." *Vanity Fair*. https://www.vanityfair.com/style/2017/02/watch-angelina-jolie-eat-bugs.

Lévi-Strauss, C. 1973. *From Honey to Ashes: Introduction to a Science of Mythology*. Translated by J. Weightman and D. Weightman. New York: Octagon.

———. 1962. *Totemism*. Translated by R. Needham. Boston: Beacon.

Lewis, C. M., and W. S. Morton. 2004. *China: Its History and Culture*. 4th ed. New York: McGraw-Hill.

Li, L. M. 1981. *China's Silk Trade: Traditional Industry in the Modern World, 1842–1937*. Cambridge, MA: Harvard University Press.

Liang, Z. S., et al. 2012. "Molecular Determinants of Scouting Behavior in Honey Bees." *Science* 335: 1225–28.

Liebhold, A., V. Mastro, and P. W. Schaefer. 1989. "Learning from the Legacy of Léopold Trouvelot." *Bulletin of the Entomological Society of America* 35: 20–22.

Lier, R.A.J. van. 1971. *Frontier Society: A Social Analysis of the History of Surinam*. Translated by M.J.L. van Ypren. The Hague: Martinus Nijhoff.

Lindsay, S. 2002. *Mount Clutter*. New York: Grove.

Lintner, J. A. 1882. *First Annual Report on the Injurious and Other Insects of the State of New York*. Albany, New York: Weed, Parsons.

Liu, Y. 2015. "Poisonous Medicine in Ancient China." In *History of Toxicology and Environmental Health: Toxicology in Antiquity*, vol. 2, edited by P. Wexler. Boston: Elsevier, 89–97.

Lochtefeld, J. G. 2002. *The Illustrated Encyclopedia of Hinduism: A–M*. New York: Rosen.

Lockwood, J. A. 2013. *Infested Mind: Why Humans Fear, Loathe, and Love Insects*. New York: Oxford University Press.

———. 2009. *Six-Legged Soldiers: Using Insects as Weapons of War*. New York: Oxford University Press.

———. 2004. *Locust: The Devastating Rise and Mysterious Disappearance of the Insect That Shaped the American Frontier*. New York: Basic Books.

Logan, W.P.D. 1953. "Mortality in the London Fog Incident, 1952." *Lancet* 261: 336–38.

Lopez, R. S. 1952. "China Silk in Europe in the Yuan Period." *Journal of the American Oriental Society* 72: 72–76.

López Binnqüist, R. C. 2003. "The Endurance of Mexican Amate Paper: Exploring Additional Dimensions to the Sustainable Development Concept." PhD thesis, University of Twente (Enschede, Netherlands).

Lorenz, E. N. 2000. "Predictability: Does the Flap of a Butterfly's Wings in Brazil Set Off a Tornado in Texas?" In *The Chaos Avant-garde: Memories of the Early Days of Chaos Theory*, edited by R. Abraham and Y. Ueda. River Edge, NJ: World Scientific, 91–94.

Lovejoy, A. O. 1936. *The Great Chain of Being: A Study of the History of an Idea*. Cambridge, MA: Harvard University Press.

Lowengard, S. 2006. *The Creation of Color in Eighteenth-Century Europe*. New York: Columbia University Press.

Lundy, M. E., and M. P. Parrella. 2015. "Crickets Are Not a Free Lunch: Protein Capture from Scalable Organic Side-Streams via High-Density Populations of *Acheta domesticus.*" *PLoS ONE* 10. DOI:10.1371/journal.pone.0118785.

L'Unione Sarda. 2008. "Most Dangerous Cheese." http://www.unica.it/pub/print.jsp?id= 6122&iso=574&is=7#casumarzu.

Ma, D. 1999. "The Great Silk Exchange." In *Pacific Centuries: Pacific and Pacific Rim History Since the Sixteenth Century,* edited by D. O. Flynn, L. Frost, and A.J.H. Latham. New York: Routledge, 38–69.

Ma, K., Y. Qiu, Y. Fu, and Q-Q Ni. 2017. "Improved shellac mediated nanoscale application drug release effect in a gastric-site drug delivery system." *RSC Advances* 7: 53401–53406.

Ma, W., R. E. Kahn, and J. A. Richt. 2009. "The Pig as a Mixing Vessel for Influenza Viruses: Human and Veterinary Implications." *Journal of Molecular and Genetic Medicine* 3: 158–66.

Maat, H. 2001. *Science Cultivating Practice: A History of Agricultural Science in the Netherlands and its Colonies, 1863–1986.* Boston: Kluwer Academic.

MacArthur, W. P. 1927. "Old Time Typhus In Britain." *Transactions of the Royal Society of Tropical Medicine and Hygiene* 20: 487–503.

MacBride, E. W. 1937. "Mendel, Morgan and Genetics." *Nature* 140: 348–50.

Mackay, M. 1861. "Some Remarks upon Shellac, with an Especial Reference to Its Present Commerical Position." *American Journal of Pharmacy and the Sciences* 9: 440–45.

MacKinnon, J. B. 2013. *The Once and Future World: Finding Wilderness in the Nature We've Made.* Boston: Houghton Mifflin Harcourt.

MacNeal, D. 2017. *Bugged: The Insects Who Rule the World and the People Obsessed with Them.* New York: St. Martin's.

Madley, B. 2016. *American Genocide: The United States and the California Indian Catastrophe, 1846–1873.* New Haven, CT: Yale University Press.

Makay, J. 1861. "Some Remarks upon Shellac, with an Especial Reference to Its Present Commercial Position." *American Journal of Pharmacy* 33: 440–45.

Makhado, R. A., M. J. Potgieter, and W. J. Luus-Powell. 2018. "Colophospermum Mopane Leaf Production and Phenology in Southern Africa's Savanna Ecosystem: A Review." *Insights of Forest Research* 2: 84–90.

Manchester City Council. 2019. "A History of Manchester Town Hall." https://www.manchester.gov.uk/info/500211/town_hall_complex/1986/a_history_of_manchester_town_hall.

Mandeville, B. 1714. *Fable of the Bees; or, Private Vices, Publick Benefits.* London: Printed for J. Roberts.

Mann, C. 2011. *1493: Uncovering the World Columbus Created.* New York: Alfred A. Knopf.

Mannheim, S. 2002. *Walt Disney and the Quest for Community.* Burlington, VT: Ashgate.

Marks, R. 1997. *Tigers, Rice, Silk, and Silt: Environment and Economy in Late Imperial South China.* New York: Cambridge University Press.

Marston, J. 1986. *The Selected Plays of John Marston.* Cambridge: Cambridge University Press.

Martin, D. 2014. *Edible: An Adventure in the World of Eating Insects and the Last Great Hope to Save the Planet*. Boston: Houghton Mifflin Harcourt.

Marx, C. 2002. *Grace Hopper: The First Woman to Program the First Computer in the United States*. New York: Rosen.

Mather, E., and J. F. Hart. 1956. "The Geography of Manure." *Land Economics* 32: 25–38.

Mawer, S. 2006. *Gregor Mendel: Planting the Seeds of Genetics*. New York: Abrams.

McCall Smith, A. 1998. *The No. 1 Ladies' Detective Agency*. New York: Pantheon.

McClellan, J. E., III. 2010 (1992). *Colonialism and Science: Saint Domingue and the Old Regime*. 2nd ed. Chicago: University of Chicago Press.

McCreery, D. 2006. "Indigo Commodity Chains in the Spanish and British Empires, 1560–1860." In *From Silver to Cocaine: Latin American Commodity Chains and the Building of the World Economy, 1500–2000*, edited by S. Topik, C. Marichal, and Z. Frank. Durham, NC: Duke University Press, 53–75.

McFadyen, R. E. 1976. "Thalidomide in America: A Brush with Tragedy." *Clio Medica* 11: 79–93.

McGrew, W. C. 2014. "The 'Other Faunivory' Revisited: Insectivory in Human and Non-human Primates and the Evolution of Human Diet." *Journal of Human Evolution* 71: 4–11.

McKenna, M. 2017. *Big Chicken: The Incredible Story of How Antibiotics Created Modern Agriculture and Changed the Way the World Eats*. Washington, DC: National Geographic.

McKibben, B. 2006 (1989). *The End of Nature*. New York: Random House.

McLaughlin, Raoul. 2016. *The Roman Empire and the Silk Routes: The Ancient World Economy and The Empires of Parthia, Central Asia and Han China*. Havertown, UK: Pen and Sword.

McLeod, C. 1987. *Hoe duur was de suiker*. Paramaribo: Vaco.

McNeill, J. R. 2010. *Mosquito Empires: Ecology and War in the Greater Caribbean, 1620–1914*. New York: Cambridge University Press.

McWilliams, J. E. 2008. *American Pests: The Losing War on Insects from Colonial Times to DDT*. New York: Columbia University Press.

Mehta, P. S., et al. 1990. "Bhopal Tragedy's Health Effects: A Review of Methyl Isocyanate Toxicity." *Journal of the American Medical Association* 264: 2781–87.

Meikle, J. L. 1995. *American Plastic: A Cultural History*. New Brunswick, NJ: Rutgers University Press.

Menzel, P., and F. D'Aluisio. 1998. *Man Eating Bugs: The Art and Science of Eating Insects*. Berkeley: Ten Speed Press.

Merchant, C. 1980. *The Death of Nature: Women, Ecology, and the Scientific Revolution*. San Francisco: Harper and Row.

Merian, M. S. 1975 (1705). *Metamorphosis insectorum Surinamensium*, edited by Helmut Decker. Amsterdam: Gerard Valck.

Merlin, C., R. J. Gegear, and S. M. Reppert. 2009. "Antennal Circadian Clocks Coordinate Sun Compass Orientation in Migratory Monarch Butterflies." *Science* 325: 1700–04.

Merrifield, M. P. 1849. *Original Treatises, Dating from the XIIth to XVIIIth Centuries, on the Arts of Painting*. 2 vols. London: John Murray.

Miall, L. C., and A. Denny. 1886. *The Structure and Life-history of the Cockroach (Periplaneta orientalis): An Introduction to the Study of Insects.* London: Lovell Reeve.

Mickens, R. E., ed. 2002. *Edward Bouchet: The First African-American Doctorate.* River Edge, NJ: World Scientific.

Millard, A. J. 2005. *America on Record: A History of Recorded Sound.* 2nd ed. New York: Cambridge University Press.

Miller, J. E. 1930. "Will Monster Insects Rule the World?" *Modern Mechanix Magazine* 5: 68–73 and 202.

Miller, N. F., and K. L. Gleason. 1994. "Fertilizer in the Identification and Analysis of Cultivated Soil." In *The Archaeology of Garden and Field,* edited by N. F. Miller and K. L. Gleason. Philadelphia: University of Pennsylvania Press, 25–43.

Mills, K., and W. B. Taylor, eds. 2006. *Colonial Spanish America: A Documentary History.* Lanham, MD: SR Books.

Millward, J. A. 2013. *The Silk Road: A Very Short Introduction.* New York: Oxford University Press.

Milne A. A. 1926. *Winnie-the-Pooh.* London: Methuen.

Mintz, S. 1985. *Sweetness and Power: The Place of Sugar in Modern History.* New York: Viking.

Misof, B., et al. 2014. "Phylogenomics resolves the timing and pattern of insect evolution." *Science* 346: 763–67.

Misra, C. S. 1928. "The Cultivation of Lac in the Plains of India (*Laccifer lacca,* Kerr)." *Bulletin of the Agricultural Research Institute, Pusa* 185: 1–115.

Mitchell, T. 2002. *Rule of Experts: Egypt, Techno-Politics, Modernity.* Berkeley: University of California Press.

Mitsuhashi, J. 2003. "Traditional Entomophagy and Medicinal Use of Insects in Japan." In *Les 'Insectes' dans la Tradition Orale,* edited by É. Motte-Florac and J. M. C. Thomas. Leuven, Belgium: Peeters, 357–65.

Moffett, M. W. 2010. *Adventures Among Ants: A Global Safari with a Cast of Trillions.* Berkeley: University of California Press.

Mohanta, J., D. G. Dey, and N. Mohanty. 2012. "Performance of Lac Insect, *Kerria lacca* Kerr in Conventional and Non-Conventional Cultivation Around Similipal Biosphere Reserve, Odisha, India." *Bioscan* 7: 237–40.

Moore, J. 2001. *An Introduction to the Invertebrates.* New York: Cambridge University Press.

Moreau de Saint Méry, M.L.É. 1798. *Description topographique, physique, civille, politique et historique de la partie française de l'isle Saint-Domingue.* 2 vols. Paris: Chez Dupont.

Morgan, E. S. 1962. *The Gentle Puritan: A Life of Ezra Stiles, 1727–1795.* Chapel Hill: University of North Carolina Press.

Morgan, T. H. 1910. "Sex-Limited Inheritance in *Drosophila.*" *Science* 32: 120–22.

Morris, B. 2004. *Insects and Human Life.* New York: Berg.

Mozzarelli, A. and S. Bettati. 2011. *Chemistry and Biochemistry of Oxygen Therapeutics from Transfusion to Artificial Blood.* Hoboken, NJ: John Wiley & Sons.

Muir, J. 1954. *The Wilderness World of John Muir,* edited by E. W. Teale. Boston: Houghton Mifflin.

Mukasonga, S. 2016. *Cockroaches.* Translated by J. Stump. New York: Archipelago.

Mukhopadhyay, B., and M. S. Muthana, eds. 1962. *A Monograph on Lac.* Bihar: Indian Lac Research Institute.

Müller, A., J. Evans, C.L.R. Payne, and R. Roberts. 2016. "Entomophagy and Power." *Journal of Insects as Food and Feed* 2: 121–36.

Müller-Maatsch, J., and C. Gras. 2016. "The 'Carmine Problem' and Potential Alternatives." In *Handbook on Natural Pigments in Food and Beverages: Industrial Applications for Improving Food Color,* edited by R. Carle and R. M. Schweiggert. Duxford, UK: Woodhead, 385–428.

Munz, T. 2016. *The Dancing Bees: Karl von Frisch and the Discovery of the Honeybee Language.* Chicago: University of Chicago Press.

———. 2005. "The Bee Battles: Karl von Frisch, Adrian Wenner and the Honey Bee Dance Language Controversy." *Journal of the History of Biology* 38: 535–70.

Murphy, M. 2008. "Chemical Regimes of Living." *Environmental History* 13: 659–703.

Mussey, R. 1981. *Transparent Furniture Finishes in New England, 1700–1820.* Ottawa: Canadian Conservation Institute.

Muthesius, A. 2003. "Silk in the Medieval World." In *The Cambridge History of Western Textiles,* 2 vols., edited by D. Jenkins. New York: Cambridge University Press, 1: 325–54.

Myerly, S. H. 1996. *British Military Spectacle: From the Napoleonic Wars Through the Crimea.* Cambridge, MA: Harvard University Press.

Myers, K. 1946. "Current Report on the Record Industry." *Notes* 3: 413.

MythBusters. 2008. "Cockroaches Survive Nuclear Explosion." https://go.discovery .com/tv-shows/mythbusters/mythbusters-database/cockroaches-survive-nuclear -explosion/.

Nash, L. 2008. "Purity and Danger: Historical Reflections on the Regulation of Environmental Pollutants." *Environmental History* 13: 651–58.

Nation, J. L. 2016. *Insect Physiology and Biochemistry.* 3rd ed. Boca Raton, FL: CRC Press.

Ndiaye, P. A. 2007. *Nylon and Bombs: DuPont and the March of Modern America.* Translated by E. Forster. Baltimore: Johns Hopkins University Press.

Negi, S. S. 1996. *Forests for Socio-economic and Rural Development in India.* New Delhi: MD Publications.

Neri, J. 2011. *The Insect and the Image: Visualizing Nature in Early Modern Europe, 1500–1700.* Minneapolis: University of Minnesota Press.

Netz, C., and S. S. Renner. 2017. "Long-Spurred *Angraecum* Orchids and Long-Tongued Sphingid Moths on Madagascar: A Time Frame for Darwin's Predicted *Xanthopan/Angraecum* Coevolution." *Biological Journal of the Linnean Society* 122: 469–78.

Nicholson, S. 1995 (1994). *Ella Fitzgerald: A Biography of the First Lady of Jazz.* New York: Da Capo.

Nobel Assembly at Karolinska Institutet. 2017. Press Release: "The Nobel Prize in Physiology or Medicine 2017." https://www.nobelprize.org/prizes/medicine/2017 /press-release/.

Noble, D. 1979. "The Chemistry of Risk: Synthesizing the Corporate Ideology of the 1980s." *Seven Days* 3: 23–26.

Norberg, R. Å. 1972. "Evolution of Flight in Insects." *Zoologica Scripta* 1: 247–50.

Nunn, J. F. 1996. *Ancient Egyptian Medicine*. Norman: University of Oklahoma Press.

O'Bannon, D. 2003. "Something Perfectly Disgusting." Disc 2 of the *Alien Quadrilogy* DVD set. 4 films. Los Angeles: Twentieth Century Fox.

Office of the United States Chief of Counsel for Prosecution of Axis Criminality. 1946. *Nazi Conspiracy and Aggression*. 8 vols. Washington, DC: U.S. Government Printing Office.

Ó Gráda, C. 2015. *Eating People Is Wrong, and Other Essays on Famine, Its Past, and Its Future*. Princeton, NJ: Princeton University Press.

Oliver, M. 1986. *Dream Work*. New York: Atlantic Monthly Press.

Oliver, R. 1975–86. "The East African Interior." In *The Cambridge History of Africa*. 8 vols., edited by J. D. Fage and R. Oliver. New York: Cambridge University Press, 3: 621–69.

Ollerton, J., R. Winfree, and S. Tarrant. 2011. "How Many Flowering Plants Are Pollinated by Animals?" *Oikos* 120: 321–26.

Orta, G. da. 1895 (1563). *Colóquios dos simples e drogas da Índia*. 2 vols. Lisbon: Imprensa Nacional.

Owen, D. 2006. *Sheetrock & Shellac: A Thinking Person's Guide to the Art and Science of Home Improvement*. New York: Simon & Schuster.

Padilla, C., and B. Anderson, eds. 2015. *Red Like No Other: How Cochineal Colored the World*. New York: Skira/Rizzoli; Santa Fe, NM: Museum of International Folk Art.

Pandey, U. B., and C. D. Nichols. 2011. "Human Disease Models in *Drosophila melanogaster* and the Role of the Fly in Therapeutic Drug Discovery." *Pharmacological Reviews* 63: 411–36.

Parikka, J. 2010. *Insect Media: An Archaeology of Animals and Technology*. Minneapolis: University of Minnesota Press.

Parry, J. H. 1966. *The Spanish Seaborne Empire*. New York: Alfred A. Knopf.

Patch, S. S. 1976. *Blue Mystery: The Story of the Hope Diamond*. Washington, DC: Smithsonian Institution Press.

Patterson, G. 2009. *The Mosquito Crusades: A History of the American Anti-Mosquito Movement from the Reed Commission to the First Earth Day*. New Brunswick, NJ: Rutgers University Press.

Payne, C.L.R., P. Scarborough, M. Rayner, and K. Nonaka. 2016. "Are Edible Insects More or Less 'Healthy' Than Commonly Consumed Meats? A Comparison Using Two Nutrient Profiling Models Developed to Combat Over- and Undernutrition." *European Journal of Clinical Nutrition* 70: 285–91.

Peck, L. L. 2005. *Consuming Splendor: Society and Culture in Seventeenth-Century England*. New York: Cambridge University Press.

Pepys, S. 2004. (1921.) *Diary of Samuel Pepys: Selected Passages*, edited by R. Le Gallienne. Mineola, NY: Dover.

Pérez-Rigueiro, J., C. Viney, J. Llorca, and M. Elices. 1998. "Silkworm Silk as an Engineering Material." *Journal of Applied Polymer Science* 70: 2439–47.

Perlstein, R. 2008. *Nixonland: The Rise of a President and the Fracturing of America*. New York: Scribner.

Petrusich, A. 2014. *Do Not Sell at Any Price: The Wild, Obsessive Hunt for the World's Rarest 78 rpm Records.* New York: Scribner.

Petty, S. 2018. *The Long Road Up from Marble Falls.* Pittsburgh: Dorance.

Petty, W. 1702. "An Apparatus to the History of the Common Practices of Dyeing, by Sir William Petty." In *The History of the Royal-Society of London for the Improving of Natural Knowledge,* edited by T. Sprat and A. Cowley. London: Printed for Richard Chiswell, 796–97.

Phillips, D. M. 2014. *Art and Architecture of Insects.* Lebanon, NH: ForeEdge.

Phipps, E. 2010. *Cochineal Red: The Art History of a Color.* New York: Metropolitan Museum of Art.

Pires, A. M., and J. A. Branco. 2010. "A Statistical Model to Explain the Mendel-Fisher Controversy." *Statistical Science* 25: 545–65.

Pliny. 1601. *The Historie of the World: Commonly Called the Natural Historie of C. Plinius Secundus.* Translated by Philemon Holland. London: Adam Islip.

———. 1855–57. *The Natural History.* Translated by John Bostock and H. T. Riley. 6 vols. London: H. G. Bohn.

Plutarch. 1916. *Lives.* Vol. 3, *Pericles and Fabius Maximus, Nicius and Crassus.* Translated by B. Perrin. Cambridge, MA: Harvard University Press.

———. 1875. *Plutarch's Lives.* Edited by Arthur Hugh Clough. Translated by J. Dryden. 5 vols. Boston: Little, Brown.

Pollens, S. 2010. *Stradivari.* New York: Cambridge University Press.

Pomeranz, K. 2000. *The Great Divergence: China, Europe, and the Making of the Modern World Economy.* Princeton, NJ: Princeton University Press.

Pomeranz, K., and S. Topik, 2006. *The World That Trade Created: Society, Culture, and the World Economy, 1400 to the Present.* Armonk, NY: M. E. Sharpe.

Price, D. A. 2009. *The Pixar Touch: The Making of a Company.* New York: Vintage.

Procopius. 1928. *History of the Wars, Books VII.36–VII.* Translated by H. B. Dewing. Cambridge, MA: Harvard University Press.

Proctor, M., P. Yeo, and A. Lack. 1996. *The Natural History of Pollination.* Portland, OR: Timber.

Prudham, W. S. 2005. *Knock on Wood: Nature as Commodity in Douglas Fir Country.* New York: Routledge.

Quezada-Euán, J.J.G., G. Nates-Parra, M. M. Maués, V. L. Imperatriz-Fonseca, and D. W. Roubik. 2018. "Economic and Cultural Values of Stingless Bees (Hymenoptera: Meliponini) Among Ethnic Groups of Tropical America." *Sociobiology* 65: 534–57.

Raffles, H. 2010. *Insectopedia.* New York: Pantheon.

———. 2007. "Jews, Lice, and History." *Public Culture* 19: 521–66.

Ramos-Elorduy, J., and P. Menzel. 1998. *Creepy Crawly Cuisine: The Gourmet Guide to Edible Insects.* Translated by N. Esteban. Rochester, VT: Park Street.

Rani, G. S. 2006. *Women in Sericulture.* New Delhi: Discovery.

Ransome, H. M. 1937. *The Sacred Bee in Ancient Times and Folklore.* Mineola, NY: Dover.

Rees, M. 1999. "Exploring Our Universe and Others." *Scientific American* 281: 78–83.

Reineccius, G. 2006. *Flavor Chemistry and Technology.* Boca Raton, FL: Taylor & Francis.

Reitsma, E. 2008. *Maria Sibylla Merian & Daughters: Women of Art and Science*. Los Angeles: J. Paul Getty Museum.

Renault, M. 2019. "Searching in Vein: A History of Artificial Blood." *Popular Science*. https://www.popsci.com/artificial-blood-history-science/.

Revkin, A. 1990. *The Burning Season: The Murder of Chico Mendes and the Fight for the Amazon Rainforest*. Boston: Houghton Mifflin.

Rijksdienst voor het Cultureel Erfgoed Ministerie van Onderwijs, Cultuur en Wetenschap. 2011. "The Iron Gall Ink Website." https://irongallink.org/.

Riley, C. V. 1871. *Sixth Annual Report on the Noxious, Beneficial and Other Insects of the State of Missouri*. Jefferson City, MO: Hegan & Carter.

Roberts, E. F. 1932. "The Clinical Application of Blow-Fly Larvae." *Scientific Monthly* 34: 531–36.

Rodríguez, L. C., and H. M. Niemeyer. 2001. "Cochineal Production: A Reviving Precolumbian Industry." *Athena Review* 2: 76–78.

Rodríguez Marín, F. 1883. *Cantos Populares Españoles Recogidos, Ordenados e Ilustrados por Francisco Rodríguez Marín*. Sevilla: Francisco Álvarez y Ca.

Romero, M. 1898. *Geographical and Statistical Notes on Mexico*. New York: G. P. Putnam's Sons.

Rosin, J., and M. Eastman. 1953. *The Road to Abundance*. New York: McGraw-Hill.

Rosselli, T. 1644. *De' secreti universali di Don Timoteo Rosselli*. Venice: Barezzi.

Rothenberg, D. 2013. *Bug Music: How Insects Gave Us Rhythm and Noise*. New York: St. Martin's.

Roxburgh, W. 1791. "Chermes Lacca, by William Roxburgh, M.D. of Samulcotta. Communicated by Patrick Russell, M.D.F.R.S." *Philosophical Transactions of the Royal Society of London* 81: 228–35.

Rücker, E., and W. T. Stearn. 1982. *Marian Sibylla Merian in Surinam*. London: Pion.

Rumold, C. U., and M. S. Aldenderfer. 2016. "Late Archaic-Early Formative Period Microbotanical Evidence for Potato at Jiskairumoko in the Titicaca Basin of Southern Peru." *Proceedings of the National Academy of Sciences* 113: 13672–77.

Russell, E. 2001. *War and Nature: Fighting Humans and Insects with Chemicals from World War I to "Silent Spring."* Cambridge: Cambridge University Press.

Sahagún, B. de. 1829–30 (1540–85). *Historia general de las cosas de la Nueva España*. 3 vols. Mexico: Imprenta del Ciudadano Alexandro Valdés.

Sainath, P. 1996. *Everybody Loves a Good Drought: Stories from India's Poorest Districts*. New York: Penguin.

Saint-Pierre, C., and O. Bingrong. 1994. "Lac Host-Trees and the Balance of Agroecosystems in South Yunnan, China." *Economic Botany* 48: 21–28.

Salaman, R. 1949. *The History and Social Influence of the Potato*. New York: Cambridge University Press.

Salazar, G. R. 1982. *Producción y comercialización de la grana cochinilla de Oaxaca y condición social de lots indígenas en la época de la colonia*. Oaxaca, MX: Talleres de Imprenta "RIOS."

Sallam, M. N. 1999. *Insect Damage: Damage on Post-harvest—Report of International*

Centre of Insect Physiology and Ecology. Rome: Food and Agriculture Organization of the United Nations, 1–34.

Saltzman, M. 1986. "Analysis of Dyes in Museum Textiles; or, You Can't Tell a Dye by Its Color." In *Textile Conservation Symposium in Honor of Pat Reevese,* edited by C. McLean and P. Connell. Los Angeles: Conservation Center, Los Angeles County Museum of Art, 27–39.

Sánchez-Bayo, F., and K.A.G. Wyckhuys. 2019. "Worldwide Decline of the Entomofauna: A Review of Its Drivers." *Biological Conservation* 232: 8–27.

Sarin, M. 1999. " 'Should I Use My Hands as Fuel?' Gender Conflicts in Joint Forest Management." In *Institutions, Relations, and Outcomes: A Framework and Case Studies for Gender-Aware Planning,* edited by Naila Kabeer and Ramya Subrahmanian. New Delhi: Kali for Women, 231–65.

Sarkar, P. C. 2002. "Applications of Lac: Past, Present and Emerging Trends." In *Recent Advances in Lac Culture,* edited by K. K. Kumar, R. Ramani, and K. K. Sharma. Ranchi: Indian Lac Research Institute, 224–30.

Schiebinger, L. 2004. *Plants and Empire: Colonial Bioprospecting in the Atlantic World.* Cambridge, MA: Harvard University Press.

———. 1989. *The Mind Has No Sex? Women in the Origins of Modern Science.* Cambridge, MA: Harvard University Press.

Schiefenhövel, W., and P. Blum. 2007. "Insects: Forgotten and Rediscovered as Food; Entomophagy Among the Eipo, Highlands of West New Guinea and in Other Traditional Societies." In *Consuming the Inedible: Neglected Dimensions of Food Choice,* edited by J. M. MacClancy, J. Henry, and H. Macbeth. New York: Berghahn, 163–76.

Schoeser, M. 2007. *Silk.* New Haven, CT: Yale University Press.

Schreiber, B. 2006. *Stop the Show! A History of Insane Incidents and Absurd Accidents in the Theater.* New York: Thunder's Mouth.

Schuh, R. T., and J. A. Slater. 1995. *True Bugs of the World (Hemiptera: Heteroptera): Classification and Natural History.* Ithaca, NY: Comstock.

Schul, J. 2000. "Carmine." In *Natural Food Colorants: Science and Technology,* edited by G. J. Lauro and F. J. Francis. New York: Marcel Dekker, 1–10.

Schur, N. W. 2013. *British English A to Zed: A Definitive Guide to the Queen's English.* New York: Skyhorse.

Schurz, W. L. 1939. *The Manila Galleon: The Romantic History of the Spanish Galleons Trading Between Manila and Acapulco.* New York: E. P. Dutton.

Schwabe, C. W. 1979. *Unmentionable Cuisine.* Charlottesville: University of Virginia Press.

Schweid, R. 1999. *The Cockroach Papers: A Compendium of History and Lore.* New York: Basic Books.

Scott, J. A. 1986. *The Butterflies of North America: A Natural History and Field Guide.* Stanford, CA: Stanford University Press.

Scott, J. C. 1998. *Seeing Like a State: How Certain Schemes to Improve the Human Condition Have Failed.* New Haven, CT: Yale University Press.

Scoville, W. C. 1952. "The Huguenots and the Diffusion of Technology—I." *Journal of Political Economy* 60, 2: 94–311.

Seeley, T. D. 2010. *Honeybee Democracy*. Princeton, NJ: Princeton University Press.

Seijas, Tatiana. 2014. *Asian Slaves in Colonial Mexico: From Chinos to Indians*. New York: Cambridge University Press.

Sen, A. 1981. *Poverty and Famines: An Essay on Entitlement and Deprivation*. New York: Oxford University Press.

Shao, Z., and F. Vollrath. 2002. "Materials: Surprising Strength of Silkworm Silk." *Nature* 418: 741.

Shapiro, F. R. 1987. "Etymology of the Computer Bug: History and Folklore." *American Speech* 62: 376–78.

Sharma, K. K. 2017. "Lac Crop Harvesting and Processing." In *Industrial Entomology*, edited by Omkar. Singapore: Springer, 181–96.

Sharma, K. K., A. K. Jaiswal, and K. K. Kumar. 2006. "Role of Lac Culture in Biodiversity Conservation: Issues at Stake and Conservation Strategy." *Current Science* 91: 894–98.

Sharma, K. K., and K. K. Kumar. 2003. "Lac Insects and Their Host-Plants." In *Potentials of Living Resources*, edited by G. Tripathi and A. Kumar. New Delhi: Discovery, 75–104.

Shaw, S. R. 2014. *Planet of the Bugs: Evolution and the Rise of Insects*. Chicago: University of Chicago Press.

Shicke, C. A. 1974. *Revolution in Sound: A Biography of the Recording Industry*. Boston: Little, Brown and Co.

Shimizu, A. 2011. "Eating Edo, Sensing Japan: Food Branding and Market Culture in Late Tokugawa Japan, 1780–1868." PhD thesis, University of Illinois at Urbana-Champaign.

Shine, I., and S. Wrobel. 1976. *Thomas Hunt Morgan: Pioneer of Genetics*. Lexington: University Press of Kentucky.

Shipek, F. C. 1981. "A Native American Adaptation to Drought: The Kumeyaay as Seen in the San Diego Mission Records, 1770–1798." *Ethnohistory* 28: 295–312.

Siegel, V. 2009. "I Kid You Not." *Disease Models & Mechanisms* 2: 5–6.

Singh, R. V. 2013. *Lac: The Wonder of Nature: A Forest Produce of Insect* Kerria lacca Kerr. Saarbrücken, Germany: LAP LAMBERT Academic.

Sleigh, C. 2003. *Ant*. London: Reaktion.

Sloane, E. 1952. *Eric Sloane's Weather Book*. New York: Hawthorn.

Smith, D. 2007. "'Le temps du plastique': The Critique of Synthetic Materials in 1950s France." *Modern & Contemporary France* 15: 135–51.

Smith, J. 1910. *Travels and Works of Captain John Smith: President of Virginia and Admiral of New England, 1580–1631*. 2 vols., edited by E. Arber. Edinburgh: John Grant.

———. 1624. *A Generall History of Virginia, New-England, and the Summer Isles*. 5 vols. London: Printed by John Dawson and John Haviland for Michael Sparkes.

Snell, G. D., and S. Reed. 1993. "William Ernest Castle, Pioneer Mammalian Geneticist." *Genetics* 133: 751–53.

So, A. 1986. *The South China Silk District: Local Historical Transformation and World-System Theory*. Albany: State University of New York Press.

Solo, R. 1955. "The New Threat of Synthetic to Natural Rubber." *Southern Economic Journal* 22: 55–64.

Song, M., et al. 2003. "Insect Vectors and Rodents Arriving in China Aboard International Transport." *Journal of Travel Medicine* 10: 241–44.

Spary, E. C. 2014. *Feeding France: New Sciences of Food: 1760–1815.* New York: Cambridge University Press.

Spear, R. J. 2005. *The Great Gypsy Moth War: The History of the First Campaign in Massachusetts to Eradicate the Gypsy Moth, 1890–1901.* Amherst: University of Massachusetts Press.

Speight, M. R., A. D. Watt, and M. D. Hunter. 1999. *Ecology of Insects.* Malden, MA: Blackwell Science.

Squires, J. E. 2002. "Artificial Blood." *Science* 295: 1002–05.

Stalker, J., and G. Parker. 1688. *A Treatise of Japaning and Varnishing: Being a Compleat Discovery of Those Arts.* Oxford: Printed by the authors.

Staller, J. E. 2010. "Ethnohistoric Sources on Foodways, Feasts, and Festivals in Mesoamerica." In *Pre-Columbian Foodways: Interdisciplinary Approaches to Food, Culture, and Markets in Ancient Mesoamerica,* edited by J. E. Staller and M. D. Carrasco. New York: Springer, 23–70.

State of Queensland, Department of Agriculture and Fisheries. 2016. "The Prickly Pear Story." https://www.daf.qld.gov.au/__data/assets/pdf_file/0014/55301/IPA-Prickly -Pear-Story-PP62.pdf.

Steinfeld, H., et al. 2006. *Livestock's Long Shadow: Environmental Issues and Options.* Rome: Food and Agriculture Organization of the United Nations.

Stephens, S. A. 2003. *Seeing Double: Intercultural Poetics in Ptolemaic Alexandria.* Berkeley: University of California Press.

Stephens, T., and R. Brynner. 2001. *Dark Remedy: The Impact of Thalidomide and Its Revival as a Vital Medicine.* New York: Basic Books.

Stevenson, R. P. 2007. *Insect Dreams.* New York: Rain Mountain.

Stoll, S. 2002. *Larding the Lean Earth: Soil and Society in Nineteenth-Century America.* New York: Hill and Wang.

Stork, N. E., J. McBroom, C. Gely, and A. J. Hamilton. 2015. "New Approaches Narrow Global Species Estimates for Beetles, Insects, and Terrestrial Arthropods." *Proceedings of the National Academy of Sciences* 112: 7519–23.

Strausfeld, N. J., and F. Hirth. 2013. "Deep Homology of Arthropod Central Complex and Vertebrate Basal Ganglia." *Science* 340: 157–61.

Stravinsky, I. 1936. *Stravinsky: An Autobiography.* New York: Simon & Schuster.

Striffler, B. F. 2005. "Life History of Goliath Birdeaters—*Theraphosa apophysis* and *Theraphosa blondi* (Araneae, Theraphosidae, Theraphosinae). *Journal of the British Tarantula Society* 21: 26–33.

Stummer, S., et al. 2010. "Application of Shellac for the Development of Probiotic Formulations." *Food Research International* 43: 1312–20.

Sturtevant, A. H. 2001. "Reminiscences of T. H. Morgan." *Genetics* 159: 1–5.

Sula, M. 2013. "South Korea." In *Street Food Around the World: An Encyclopedia of Food and Culture,* edited by B. Kraig and C. T. Sen. Santa Barbara, CA: ABC-CLIO, 316–22.

Sun, W., et al. 2012. "Phylogeny and Evolutionary History of the Silkworm." *Science China Life Sciences* 55: 483–96.

Suter, A. F. 1911. "Technical Notes on Lac: A Paper Read by A. F. Suter Before the Paint and Varnish Society of London." *Paint, Oil and Drug Review* 51: 36–38.

Sutter, P. 2007. "Nature's Agents or Agents of Empire? Entomological Workers and Environmental Change During the Construction of the Panama Canal." *Isis* 98: 724–54.

Tavernier, J.-B. 1925. *Travels in India*. Translated by V. Ball. 2 vols. 2nd ed. London: Oxford University Press.

Taylor, A. 2014. "Bhopal: The World's Worst Industrial Disaster, 30 Years Later." *Atlantic*. https://www.theatlantic.com/photo/2014/12/bhopal-the-worlds-worst-industrial-disaster-30-years-later/100864/.

Taylor, K. 2019. "Eating Insects Will Soon Go Mainstream as Bug Protein Is Set to Explode into an $8 Billion Business." *Business Insider*. https://www.businessinsider.com/eating-insects-set-to-become-8-billion-business-barclays-2019-6.

Tedlock, J., ed. and trans., 1996. *Popol Vuh: The Mayan Book of the Dawn of Life*. New York: Simon & Schuster.

Thiéry de Menonville, N.-J. 1787. *Traité de la culture du nopal et de l'éducation de la cochenille dans les colonies françaises de l'Amérique, précédé d'un voyage à Guaxaca*. 2 vols. Cap-Français: Chez la veuve Herbault, Libraire de Monseigneur le Général & du Cercle des Philadelphes.

Thirsk, J. 1997. *Alternative Agriculture: A History from the Black Death to the Present Day*. New York: Oxford University Press.

Thomas, M. C., et al. 2019. "Mediterranean Fruit Fly, *Ceratitis capitate* (Wiedmann) (Insecta: Diptera: Tephritidae) *University of Florida Institute of Food and Agricultural Sciences Extension*. https://edis.ifas.ufl.edu/pdffiles/IN/IN37100.pdf.

Thompson, E. 1995. "Machines, Music, and the Quest for Fidelity: Marketing the Edison Phonograph in America, 1877–1925." *Musical Quarterly* 79: 131–71.

Thoreau, H. D. 1873 (1849). *A Week on the Concord and Merrimack Rivers*. Boston: James R. Osgood.

Thorpe, W. H. 1954. "Life and Senses of the Honey Bee." *Nature* 174: 897.

Todd, K. 2007. *Chrysalis: Maria Sibylla Merian and the Secrets of Metamorphosis*. Orlando, FL: Harcourt.

Toussaint-Samat, M. 2009. *A History of Food*. Translated by Anthea Bell. 2nd ed. Malden, MA: Wiley-Blackwell.

Travis, A. S. 2007. "Anilines: Historical Background." In *The Chemistry of Anilines*, edited by Z. Rappoport. Hoboken, NJ: John Wiley & Sons.

Trouvelot, É. L. 1882. *The Trouvelot Astronomical Drawings Manual*. New York: Charles Scribner's Sons.

———. 1867. "The American Silkworm." *American Naturalist* 1: 30–38.

Tsutsui, W. M. 2007. "Looking Straight at Them! Understanding the Big Bug Movies of the 1950s." *Environmental History* 12: 237–53.

Turner, C. H. 1892. "A Few Characteristics of the Avian Brain." *Science* 19: 16–17.

Turner, R. L. 2008 (1962–66). *A Comparative Dictionary of the Indo-Aryan Languages.* Delhi: Motilal Banarsidass.

Ujváry, I. 1999. "Nicotine and Other Insecticidal Alkaloids." In *Nicotinoid Insecticides and the Nicotinic Acetylcholine Receptor,* edited by I. Yamamoto and J. E. Casida. Tokyo: Springer-Verlag, 29–69.

UMass Extension. 2015. "Distinguish Between Eastern Tent and Gypsy Moth Caterpillars." https://ag.umass.edu/sites/ag.umass.edu/files/fact-sheets/pdf/tent_and_gypsy_moth_caterpillars.pdf.

United Nations Department of Economic and Social Affairs. 2015. "World Population Projected to Reach 9.7 Billion by 2050." http://www.un.org/en/development/desa/news/population/2015-report.html.

United Nations News Center. 2015. "UN Environment Chief Warns of 'Tsunami' of E-waste at Conference on Chemical Treaties." https://www.un.org/sustainabledevelopment/blog/2015/05/un-environment-chief-warns-of-tsunami-of-e-waste-at-conference-on-chemical-treaties/.

United States Department of Agriculture. 1869. *Report of the Commissioner of Agriculture for the Year 1868.* Washington, DC: U.S. Government Printing Office.

United States Department of Agriculture Economic Research Service. 2018. "Sugar and Sweeteners." https://www.ers.usda.gov/topics/crops/sugar-sweeteners/.

United States Food and Drug Administration. 2018. *Defect Levels Handbook.* https://www.fda.gov/food/guidanceregulation/guidancedocumentsregulatoryinformation/sanitationtransportation/ucm056174.htm.

Usher, M. B., and M. Edwards. 1984. "A Dipteran from South of the Antarctic Circle: *Belgica antarctica* (Chironomidae) with a Description of Its Larva." *Biological Journal of the Linnean Society* 23: 19–31.

Vainker, S. 2004. *Chinese Silk: A Cultural History.* London: British Museum Press.

van Huis, A., et al. 2013. *Edible Insects: Future Prospects for Food and Feed Security.* Rome: Food and Agriculture Organization of the United Nations.

Van Lier, R.A.J. 1971. *Frontier Society: A Social Analysis of the History of Surinam.* The Hague: Martinus Nijhoff.

Van Linschoten, J. H. 1885. *The Voyage of John Huygen van Linschoten to the East Indies,* edited by A. C. Burnell and P. A. Tiele. London: Whiting.

Varshney, R. K. 1970. *Lac Literature: A Bibliography of Lac Insects & Shellac.* Calcutta: Shellac Export Promotion Council.

Verosub, K. L., and J. Lippman. 2008. "Global Impacts of the 1600 Eruption of Peru's Huaynaputina Volcano." *EOS, Transaction of the American Geophysical Union* 89: 141–48.

Vogel, S. A. 2008. "From 'The Dose Makes the Poison' to 'The Timing Makes the Poison': Conceptualizing Risk in the Synthetic Age." *Environmental History* 13: 667–73.

Voloshin, M. 2002. "The Preservation and Storage of Historical 78 rpm Recorded Discs." *Music Reference Services Quarterly* 8: 39–43.

Voltaire. 1876–78. *Oeuvres complètes de Voltaire.* 13 vols. Paris: Firmin-Didot.

Von Frisch, K. R. 1967. *The Dance Language and Orientation of Bees.* Translated by L. Chadwick. Cambridge, MA: Harvard University Press.

———. 1954 (1927). *The Dancing Bees: An Account of the Life and Senses of the Honey Bee.* Translated by Dora Ilse. London: Methuen.

———. 1936. *Du und das Leben: Eine moderne Biologie für Jedermann.* Berlin: Ullstein.

Voosen, Paul. 2018. "Outer Space May Have Just Gotten a Bit Closer." *Science.* https://www.sciencemag.org/news/2018/07/outer-space-may-have-just-gotten-bit-closer.

Waley, P. 2000. "What's a River Without Fish? Symbol, Space and Ecosystem in the Waterways of Japan." In *Animal Spaces, Beastly Places: New Geographies of Human-Animal Relations,* edited by C. Philo and C. Wilbert. New York: Routledge, 159–81.

Wallace, A. R. 1867. "Creation by Law." *Quarterly Journal of Science* 4: 471–88.

Wallace, R. 1952. "First It Said 'Mary': The Phonograph Is Celebrating Its 75th Year." *Life,* 87–102.

Wang, A., et al. 2018. "A Suspect Screening Method for Characterizing Multiple Chemical Exposures Among a Demographically Diverse Population of Pregnant Women in San Francisco." *Environmental Health Perspectives* 126. https://doi.org/10.1289/EHP2920.

Wangler, M. F., S. Yamamoto, and H. J. Bellen. 2015. "Fruit Flies in Biomedical Research." *Genetics* 199: 639–53.

Ward, J. E., ed. 2008. *The Book of Odes (Shijing).* Translated by J. Legge. Morrisville, NC: Lulu Books.

Wargo, J. 1998. *Our Children's Toxic Legacy: How Science and Law Fail to Protect Us from Pesticides.* New Haven, CT: Yale University Press.

Washington, G. 2003. *Washington on Washington,* edited by P. M. Zall. Lexington: University Press of Kentucky.

Watt, G. 1908. *The Commercial Products of India: Being an Abridgement of "The Dictionary of the Economic Products of India."* London: John Murray.

———. 1905. "The Lac Industry of India." *Pharmaceutical Journal* 75: 650–52.

Waugh, F. W. 1916. "Iroquois Foods and Food Preparation." In *Canadian Department of Mines, Geological Survey,* Mem. 86, no. 12, *Anthropological Series.* Ottawa: Government Printing Bureau, 138–39.

Webb, M. 2000. *Lacquer Technology and Conservation: A Comprehensive Guide to the Technology and Conservation of Asian and European Lacquer.* Woburn, MA: Butterworth-Heinemann.

Weber, M. 1946 (1918). "Science as a Vocation." In *From Max Weber: Essays in Sociology,* edited by H. H. Gerth and C. W. Mills. New York: Oxford University Press, 129–56.

Webster, F. M. 1897. "The Periodical *Cicada septendecim,* or So-called Seventeen Year Locust, in Ohio." *Ohio Agricultural Experiment Station* 87: 37–68.

Wehner, R. 2016. "Early Ant Trajectories: Spatial Behaviour Before Behaviourism." *Journal of Comparative Physiology A* 202: 247–66.

Weiner, J. 1999. *Time, Love, Memory: A Great Biologist and His Quest for the Origins of Behavior.* New York: Alfred A. Knopf.

Weiss, M. R. 1991. "Floral Colour Changes as Cues for Pollinators." *Nature* 354: 227–29.

Westwood, J. O. 1883. "On the Probable Number of Species of Insects in the Creation; Together with Descriptions of Several Minute Hymenoptera." *Magazine of Natural*

History, and Journal of Zoology, Botany, Mineralogy, Geology, and Meteorology 6: 116–123.

Wexler, P. 2018. *Toxicology in Antiquity.* 2nd ed. London: Academic.

Wheeler, W. M. 1992. "Social Life Among the Insects." *Scientific Monthly* 15: 385–404.

White House, Office of the Press Secretary. 2010. https://obamawhitehouse.archives.gov /the-press-office/2010/11/03/press-conference-president.

Whitfield, S. 2018. *Silk, Slaves, and Stupas: Material Culture of the Silk Road.* Berkeley: University of California Press.

Whitman, W. 2007 (1855). *Leaves of Grass.* Mineola, NY: Dover.

Wigglesworth, V. B. 1942. *The Principles of Insect Physiology.* 2nd ed. London: Methuen.

Wild Colors from Nature website (accessed on January, 30 2013). http://www.wildflavors .com/?page=cochineal_carmine.

Wilde, O. 1894 (1893). *A Woman of No Importance.* London: John Lane.

———. 1891. *The Picture of Dorian Gray.* London: Ward, Lock.

Wilkes, J., and J. Adlard, eds. 1810. *Encyclopaedia Londinensis; or, Universal Dictionary of Arts, Sciences, and Literature.* 24 vols. London: J. Adlard Printer.

Williams, L. O. 1970. "Jalap or Veracruz Jalap and Its Allies." *Economic Botany* 24: 399–401.

Wilson, D., and A. Rhodes. 2016. "Dream of Wild Health: As They Plant a Meadow for Pollinators, Native American Teenagers Are Building Their Relationship to the Earth." *Minnesota Conservation Volunteer.* https://www.dnr.state.mn.us/mcv magazine/issues/2016/jul-aug/dream-of-wild-health.html.

Wilson, E. O. 1990. "First Word." *Omni* 12: 6.

———. 1984. *Biophilia.* Cambridge, MA: Harvard University Press.

Winegard, T. C. 2019. *The Mosquito: A Human History of Our Deadliest Predator.* New York: Dutton.

Wise, W. 1968. *Killer Smog: The World's Worst Air Pollution Disaster.* Chicago: Rand McNally.

Witmer, P. 2018. "Exactly How Nutritious Was Justin Timberlake's Bug Buffet?" *Noisey VICE.* https://noisey.vice.com/en_ca/article/3k5yv5/exactly-how-nutritious-was -justin-timberlakes-bug-buffet.

Wöhler, F. 1828. "Ueber künstliche Bildung des Harnstoffs." *Annalen der Physik und Chemie.* 88: 253–56.

Wood, F. 2002. *The Silk Road: Two Thousand Years in the Heart of Asia.* Berkeley: University of California Press.

Woodgate, J. L., J. C. Makinson, K. S. Lim, A. M. Reynolds, and L. Chittka. 2017. "Continuous Radar Tracking Illustrates the Development of Multi-destination Routes of Bumblebees." *Scientific Reports* 7: 1–15.

World Health Organization, 2020. "Global Health Observatory (GHO) Data." https:// www.who.int/gho/malaria/epidemic/deaths/en/.

———. 2016. "Zika Situation Report." http://www.who.int/emergencies/zika-virus /situation-report/16-june-2016/en/.

Worster, D. 1977. *Nature's Economy: A History of Ecological Ideas.* New York: Cambridge University Press.

Wright, S. E. 2015. *Tying Heritage Featherwing Streamers*. Mechanicsburg, PA: Stackpole.

Wrolstad, R. E., and C. A. Culver. 2012. "Alternatives to Those Artificial FD&C Food Colorants." *Annual Review of Food Science and Technology* 3: 59–77.

Xue, Y. 2005. "'Treasure Nightsoil as If It Were Gold': Economic and Ecological Links Between Urban and Rural Areas in Late Imperial Jiangnan." *Late Imperial China* 26: 41–71.

Yale University. 2017. "Yale to Change Calhoun College's Name to Honor Grace Murray Hopper." *YaleNews*. http://news.yale.edu/2017/02/11/yale-change-calhoun-college -s-name-honor-grace-murray-hopper-0.

Yang, J., Y. Yang, W. Wu, J. Zhao, and L. Jiang. 2014. "Evidence of Polyethylene Biodegradation by Bacterial Strains from the Guts of Plastic-Eating Waxworms." *Environmental Science & Technology* 48: 13776–84.

Yen, A. L. 2009. "Entomophagy and Insect Conservation: Some Thoughts for Digestion." *Journal of Insect Conservation* 13: 667–70.

Yimin, H. 1995. "Sichuan Province Reforms Under Governor-General Xiliang, 1903–1907." In *China, 1895–1912: State-Sponsored Reforms and China's Late-Qing Revolution*, edited and translated by D. R. Reynolds. Armonk, NY: M. E. Sharpe, 136–56.

Yu, Y. 1967. *Trade and Expansion in Han China: A Study in the Structure of Sino-Barbarian Economic Relations*. Berkeley: University of California Press.

Yule, Col. H., and A. C. Burnell. 1903. *Hobson-Jobson: A Glossary of Colloquial Anglo-Indian Words and Phrases, and of Kindred Terms, Etymological, Historical, Geographical and Discursive*. London: J. Murray.

Zhang, L., et al. 2018. "A Nonrestrictive, Weight Loss Diet Focused on Fiber and Lean Protein Increase." *Nutrition* 54: 12–18.

Zion Market Research. 2018. "Insecticides Market to Head North With An Estimated Value of USD 3.48 Billion in 2021." https://www.zionmarketresearch.com/news /global-insecticides-market.

Zinsser, H. 1935. *Rats, Lice and History*. Boston: Little, Brown.

Zuk, M. 2013. *Paleofantasy: What Evolution Really Tells Us about Sex, Diet, and How We Live*. New York: W. W. Norton.

INDEX

Page references in *italics* refer to illustrations.

Abdülmecid I , Ottoman sultan, 13, 14, 38
Adams, Roger, 97
Adivasis, 43–4
Aesop, 33
Africa, traditions of entomophagy in, 162–4
Agnew, Spiro, 43
agriculture. *See* food production
agritecture, 169
Aikins, R. C., 148
Akbar I, Mughal emperor, 7, 41
alcoholic beverages, honey in, 36
Aleppo galls, 37
Alien, 33
almond tree *(Prunus dulcis),* 145
Amazon basin, 27, 174
American Association for the Advancement of Science, 6, 129
American Naturalist, 74–5
American Revolution, 13–14
Andes: cochineal dye in, 85–6, 92; potato cultivation in, 156
Ando, Momofuku, 169
Anopheles mosquitoes, 24, 29
"Ant and the Grasshopper, A" (Aesop), 33
Antarctica, 7
Anthropocene, 177
Antonio's Revenge (Marston), 37
ants, 27, 139, 140; behavioral patterns in, 25; *A Bug's Life* and, 33–4; as exemplars of diligent labor, 16–17;

as food, 153, 154, 159; human behavioral patterns analogous to, 25–6; leaf-cutter (genera *Atta* and *Acromyrmex*), 111, 197n; termite-hunting *(Megaponera analis),* 26; velvety tree *(Liometopum apiculatum),* 164–5
aphids, 3, 30, 86, 149
Apidae, 146
apocalypse, 32. *See also* nuclear apocalypse
Arce, Manny, 164
Aristotle, 20
arthropods, 14, 18; current epoch as age of, 177; "evil," in movies, 32–3
Atomic Age, 109–10
Aus dem Leben der Bienen (von Frisch), 141–2
Australia, 154; huge cockroaches of, 111; *Opuntia* cacti introduced in, 92; Yolngu people of, 35–6, 146
Ayurveda, 40
Aztecs, 35, 81, 86, 89, 165

Bakelite, 99n
Bancroft, Edward, 44
Banda, Hastings Kamuzu, 164
Baroque painting, cochineal dye in, 90
Barthes, Roland, 99
Bateman, James, 134
Bateson, William, 128, 129

battlefield wounds, larvae in treatment of, 30

beauty products: shellac in, 54, 55, 108; silkworm cocoons in, 106

Beck, Paul, 106

Becket, Thomas, 15

bedbugs, 3, 30

Beecher, Catharine Esther, 155

beef. *See* meat

beehives, 16, 34, 35–6, 145, 147–9; bees' communication about food sources near, 141–2; economic theorists and, 17

beekeeping, or apiculture, 34–5, 185n; Colony Collapse Disorder and, 147–9; migratory, 144–6

bees, 139, 159, 177, 182n; as exemplars of diligent labor, 16–17; imagining world without, 146–7; *Melipona,* raised by Aztecs, 35, 185n; native to North America, 146; as pollinators, 4, 137–8, 139, 141–2, 144–6, 185n; sugarbag *(Tetragonula carbonaria),* in Australia, 35–6; sweat, 147. *See also* honeybees

beeswax, 35, 44

beetles, 3, 4, 22, 26, 27, 30, 110, 137, 149, 159, 187n; eating larvae of, 154–6

Belgica antarctica, 7

Bellarmine, Robert, 15

Benét, Stephen Vincent, 115

Bengal famine (1943), 170

Benson, Ezra Taft, 101

beondegi (Korean street food), 161

Beowulf, 36

Berggren Åsa, 169–70

Berliner, Emile, 47, 48

Bernofsky, Susan, 112–13

Berzelius, Jöns Jacob, 96

Bhopal, India, industrial disaster (1984), 102

Big Pine Paiute Tribe, 166

biodegradation, 103, 108

Biophilia (Wilson), 22

biopiracy, 190–1n; Thiéry's theft of cochineal bugs, 5, 78–85, 91–2

biosphere, 177n; importance of insects vs. humans to, 9–10; pollination vital to, 146

Bismarck, Otto von, 18

Blair, David, 19n

Blake, William, 131–2

Blass, H. Bentley, 110

blood: artificial, 104; stage, cochineal as, 91

bloodsuckers, 28–9

Bodenheimer, Friedrich Shimon, 172

boll weevil *(Anthonomus grandis),* 30

Bookchin, Murray, 195n

Bouchet, Edward A., 141

Bourdain, Anthony, 155

Boveri, Theodor, 129

bowling alleys, 54

Boxer, Charles Ralph, 22

Bradbury, Ray, 181n

Brazil, 106; mead in, 36; melodic rituals of insects in rain forest of, 174

brazilwood *(Paubrasilia echinata),* 93

Breslin, Catherine, 110

Bridges, Calvin, 129

British Central Africa (Johnston), 31–2

British East India Company, 42–3, 92

Brues, Charles Thomas, 167

Brulley, A. J., 85

Bryant, Edwin, 166

Bryson, Bill, 187n

Bucareli y Ursúa, Antonio María de, 82

bug concerts *(mushi-kiki),* 174

"bug in the system," 31, *31*

Bug Music (Rothenberg), 175

Bug's Life, A, 33

"bugs," use of term, 30–1

bumblebees, 146, 175

butterflies, 6, 18–19, 22, 24, 150, 159, 177; monarch *(Danaus plexippus),* 6, 18–19, 123n, 175–6; moths vs., 56n;

as pollinators, 4, 137; wide-ranging effects of, 6, 178, 181n; Xerces blue *(Glaucopsyche xerces)*, 150

BuzzBuilding, 169

Byzantine Empire, sericulture in, 66, 188n

Calonius, Erik, 112

Campbell, George, 135

campeche wood, or *palo de campeche (Haematoxylum campechianum)*, 93

Canary Islands, cochineal in, 93, 193n, 194n

cancer, 94, 104, 132

Candide (Voltaire), 90

Caravaggio, Michelangelo Merisi da, 90

carmine. *See* cochineal dye

Carothers, Wallace, 77

carrion insects, 30

Carson, Rachel, 7, 101–2, *102*, 123–4, 149, 150

caveman, or paleolithic, diet, 159–60

cave paintings, 34

Central Valley, California, 144–6

chapulines (toasted grasshoppers), 164

Charles V, Holy Roman Emperor, 86, 89

Charles IX, king of France, 74

Charter of the Forest, English, 35

cheese flies *(Piophila casei)*, 158–9

Chernobyl disaster (1986), 197n

Chetverikov, Sergei, 27

Chiang Yee (Jiang Yi), 76

China, 29, 94, 100; aesthetic appeal of "silkworm-moth eyebrows" in, 56–7; airborne transit of cockroaches to, 112; manure recycling in, 71–2; shellac and, 41, 42, 46, 108, 186n; silk production in, 5, 58–9, 61–2, 63–9, 71–2, 106, 189–90n; trade with, 42, 46, 67–9, 89

china poblana silk dress, 67, 68–9

cholera, 106

chromosomes, 133; of fruit flies, Morgan's experiments on, 8, 120, 122, 124–5,

128–9; Mendel's laws of inheritance and, 129

cicadas, 30, 86; Meleager of Gadara's ode to, 173–4; sounds of, 174, 175

circadian rhythm, *Drosophila* and, 133

Civil War, 29, 120, 121, 150

Clavijero, Francisco Javier, 167

climate change, 115, 168, 176

clothing, moth larvae and, 26

Cloudsley-Thompson, John, 28–9

Cobo, Bernabé, 87

cochineal, Polish *(Porphyrophora polonica)*, 90

cochineal dye, 4, 8, 13–14, 45, 78–95, 105, 168, 177; carminic acid in, 79n, 87, 94, 109; costliness of, 78–9, 93, 94; decline and revival of, 38, 93–5, 97, 99, 100, 104, 107, 109, 116; European red dyes prior to, 90; failed attempts at production of, 92, 104; as food colorant, 94, 155; oigin and history of, 85–6, 91; painters' use of, 90; preparation of, 79n, 88; synthetic substitutes for, 8, 104; in textiles, 92–3

cochineal farmers *(nopaleros)*, 88

cochineal insects *(Dactylopius coccus)*, 4, 14, 104; British and Australian cultivators of, 92; harvesting, 88; host plants for, 40, 45 *(see also* nopal, or prickly pear, cactus); nurturing, 87–8; Thiéry's smuggling of, 5, 78–85, 91–2

cockchafer grubs, sautéed, 155–6

cockroaches, 3, 109–15, 116, 140, 162; "La Cucaracha" and, 113–14, 197n; diversity in size of, 111; eaten in amusement park contest, 152–3; genocidal rhetoric and, 114; global disperson of, 112; with immunity to new insecticides, 114–15; Madagascar hissing *(Gromphadorhina portentosa)*, 152, *152*–3; Marquis's

cockroaches *(continued)*
  "Archy" and, 113, *113*; nuclear
    apocalypse and, 109–10, 111, 114, 115
Codex Mendoza, 35, 86, 191–2n
Codex Sinaiticus, 38
coevolution, 135, 137
coffee beans, insect fragments and, 168
Colony Collapse Disorder (CCD), 147–9
color, as visual signal for pollinators, 138,
    139
colorants, 90, 192n; lac, or shellac, as, 41,
    44, 47; Native American cloth-
    coloring techniques and, 92–3. *See
    also* cochineal dye
Columbian Exchange, 89
Columbia University, Fly Room at, 8–9,
    119–20, 122, 124–6, 128–9, 132
Columbus, Christopher, 89, 90
Columella, Lucius, 71
"commensal," fruit fly as, 130
"Common Fruit Fly, The" (Lindsay), 132
confections, shellac glazes and, 107
Confucius, 61, 64
Congress, U.S., 100–1, 102, 104
conquistadors, 35, 81, 86
Coon, Julius M., 101
Cordain, Loren, 159
Cornetz, Victor, 140
Cortés, Hernán, 81, 86, 89
"cosmopolitan organism," fruit fly as, 130
cotton, 22, 42, 60, 63, 80, 97, 98, 105, 106;
    boll weevil and, 30
Crandall, Elizabeth Brownell, 48–9
Crassus, Marcus Licinius, 62, 188n
Crawford, Mabel Sharman, 58
crickets, 4, 9; Jiminy Cricket, *33, 34*;
    new products fortified with, 160–1;
    raising as food, 153–4, 170; sounds
    of, 174–5
crime scene investigations, 30
Crosby, Alfred W., 89–90
crossbreeding, 126
Crutzen, Paul, 177n

Cuban, Mark, 9
"Cucaracha, La," 113–14, 197n
Cuevas de la Araña, Valencia, Spain, 34
Curculionidae, 27

Daguerre, Louis-Jacques-Mandé, 46
D'Aluisio, Faith, 154
Dancing Bees, The (von Frisch), 141–2
Darwin, Charles, 10, 128, 134–7, 151, 178
Charles Darwin Research Station,
    Galápagos Islands, 130
Davis, Mike, 69
"de-bugging," 31
Declaration of Independence, 38, 46
decomposition of organic matter, 177
Defoe, Daniel, 147
Delaney, James J., 104
dentistry, shellac in, 108
Descent of Man, The (Darwin), 10
DeVries, Philip, 136–7
Dicke, Marcel, 168
Dickinson, Emily, 10
Diet for a Small Planet (Lappé), 168
Dillard, Annie, 6–7
Diogenes the Cynic, 20
Discovery Channel, 111
diseases, insect-borne, 3, 28–9, 173
Walt Disney Company: *A Bug's Life,* 33–4;
    "House of the Future" display, *98,*
    98–9; *Pinocchio, 33, 34*
Dodds, Warren "Baby," 50
"domestication," as term, 58
Don Bugito, the Prehispanic Snackeria,
    San Francisco, 165
Donne, John, 15–16
doodlebugs, sautéed larvae of, 155–6
dragonflies, 16, *17*
Du Bois, Christine M., 171
DuPont Chemical Company, 7, 77, 100
DuPuy, William Atherton, 176
Dürer, Albrecht, 20, 110
Duryodhana, 39–40, 54
Dutch West India Company, 22

Eastman, Max, 97, 194n
Eaton, Stanley Boyd, 160
Ebers Papyrus, 100
Edison, Thomas Alva, 30–1, 48, 53
Edwardes, Charles, 93
Einstein, Albert, 144, 146
Eisenhower, Dwight, 101
Elagabalus, Roman emperor, 63
electronics industry, shellac in, 108
El Niño–Southern Oscillation
    (ENSO), 69
embalming, shellac in, 108
Encyclopaedia Londinensis (Wilkes), 96
Enterprise, Ala., boll weevil and, 30
entomophagy (eating of insects), 5, 9,
    152–72, 177; advocacy organizations
    and, 161; in amusement park contest,
    152–3; aversion in Western societies
    to, 155, 167, 172; challenges to global
    food system in twenty-first century
    and, 168–72; cricket-flour start-
    ups and, 160–1; culinary tourism
    and, 154–5; cultures with venerable
    traditions of, 161–7; ecological
    and nutritional benefits of, 153–4;
    high-profile promotional efforts
    and, 154; insect fragments in foods
    and, 168; interstellar exploration
    and, 171; nineteenth-century recipe
    books and, 155–6; origin of term,
    153; radical alterations of food
    preferences and, 156–8; unequal
    access to food and, 170
environmental advocacy, 149–50
environmental toxicology, 99, 100–3
Errington, Fred, 169
escamoles (Mexican insect delicacy), 164–5
Eugenides, Jeffrey, 61
European Union, 107, 159
extinctions, 10, 115–16, 150, 176

Fable of the Bees (Mandeville), 17
Fabre, Jean-Henri, 178

Fearnley-Whittingstall, Hugh, 168
Ferchault de Réaumur, René Antoine,
    155n
Fernald, Maria Elizabeth, 75
Fisher, M.F.K., 156
Fisher, Ronald, 127n
Fitzgerald, Ella, 13, 14, 38
fleas, 28, 29; Donne's "The Flea," 15–16
flies, 4, 28, 137; battlefield wounds and
    larvae of, 30; distinctive pitches
    emanating from, 173; petroleum
    (Helaeomyia petrolei), 27
Flight Behavior (Kingsolver), 175–6
"Flight of the Bumblebee" (Rimsky-
    Korsakov), 175
Florentine Codex, 89, 165, 192n
"Fly, The" (Blake), 131–2
fly-fishing, 107
"fly on the wall" fantasy, 176
food: insects as (*see* entomophagy);
    shellac sprayed on, 107–8
Food, Drug, and Cosmetic Act of 1938,
    104
food additives, synthetic, 99, 104, 160
Food and Drug Administration (FDA),
    94, 107, 168
food colorants, 94, 104, 155
food production: breeding to achieve
    desirable traits and, 126; insects
    central to health of, 177; manure
    and compost in, 71–2; pesticide-
    and herbicide-dependent, 116,
    176; pollination and, 137, 144–9;
    potato propagation and, 156–71;
    reliance on exotic honeybees and,
    145–6; shellac's uses in, 55, 107–8; in
    Synthetic Age, 97–8
Foote, Samuel, 43
Fortin, Ernest A., 146
Foucher d'Obsonville, 155n
Francis, Clive, 91
Franklin, Benjamin, 157
French Revolution, 85

Freud, Sigmund, 142–3
Frisch, Karl von, 138, 141–4, *143*
fruit flies *(Drosophila melanogaster)*,
    4, 8–9, 119–33, *125*; as both
    "commensal" and "cosmopolitan
    organism," 130; favorable attributes
    of, for scientific study, 125; "Fly
    Room" and, 8–9, 119–20, 122, 124–6,
    128–9, 132; genetic resemblances
    of human body and, 9; as getaway
    artists, 130; launched into space,
    171; Lindsay's poem about, 132;
    migration of, to United States,
    129–30; Morgan's experiments on, 8,
    120, 122, 124–5, 128–9; Nobel Prizes
    for work with, 9, 120, 122–3, 133;
    Palin's derogatory remark about,
    133; self-fertilization of, 128–9; sleep
    cycles and, 132–3
fruit flies, Mediterranean *(Ceratitis
    capitata)*, 119n, 198n
Führer, Wilhelm, 144
Fujikura, Tatsuro, 169

"Gadfly Said to the Fly, The" (Grieg), 175
Galeano, Eduardo, 81
Gallatitlán pueblo, Mexico, 83–4
Garcia, Anna, 164
Garcia da Orta, 44
Garibaldi, Giuseppe, 120
Gates, Bill, 9, 45
Gauguin, Paul, 90
Gaul, Albro T., 175n
Gaye, Marvin, 102–3
genetics. *See* chromosomes; inheritance
"genetics," first use of term, 128
Genghis Khan, 67
genocidal rhetoric, insect epithets in, 32,
    114
Gewertz, Deborah, 169
"ghost acreage," 103–4
Glass, H. Bentley, 110
gnats, 30, 77, 133

goliath birdeaters *(Theraphosa
    blondi)*, 23
Gómez de Cervantes, Gonzalo, 90
Gore, Al, 103
Gould, Stephen Jay, 177
Graduate, The, 99
Gramophone, 7
gramophones, 53–4
grasshoppers, 4; *A Bug's Life* and, 33–4; as
    food, 153, 154, 164, 165–6
Great Chain of Being, 28
Greece, ancient, 20, 96, 100; beekeeping
    in, 34; regard for insect songs in,
    173–4
Greek, derivation of words "insect" and
    "arthropod" and, 18
greenhouse gases, raising livestock vs.
    edible insects and, 153
Grieg, Edvard, 7, 175
griffinflies *(Meganeura monyi)*, 26
grubs, as food, 153, 154, 155–6
Grupen, Claus, 111
Guarneri, Giuseppe, 42
gypsy moths *(Lymantria dispar)*, 75, 92,
    190n

habitat destruction, 150, 176
Haldane, J.B.S., 27
Hall, Jeffrey C., 9, 132–3
Hanna-Barbera, 98
Hanson, Thor, 146–7
Hawaiian wēkiu bug *(Nysius wekiuicola)*,
    27, *28*
Hayworth, Rita, 54
Hemiptera, 30
hemoglobin, 104
Henry II, king of England, 15
Hesse-Honegger, Cornelia, 197n
Himmler, Heinrich, 32
Hindenburg, Paul von, 143
Hiroshima bombing (1945), 109–10, 111
Historia animalium (Aristotle), 20
Historia naturalis (Pliny the Elder), 20

Hitler, Adolf, 143

Hoffman, Walter J., 165–6

Hofmann, August Wilhelm von, 93

Holland, Philemon, 20

Holocaust, 32

Holt, Vincent M., 155, 172

hominid foragers, diet of, 159–60

honey, 7–8, 34–7, 38, 42, 44, 86, 159;
    decline of, 36–7; earliest known
    depictions of hunting for, 34; in
    mead, 36; Yolngu "sugarbag" and,
    35–6

honeybees, 18, 34–6, 192n; Carniolan
    (Apis mellifera carnica), 141–2;
    Colony Collapse Disorder
    and, 147–9; colors and patterns
    perceived by, 139; European *(Apis
    mellifera)*, 145–6; human behavioral
    patterns analogous to, 25–6;
    migratory beekeepers and, 144–6;
    Turner's experiments on, 139–40,
    141; von Frisch's description of
    communication rituals of, 141–2

"honey," metaphorical uses of word, 37

Hooke, Robert, 173

Hooker, Joseph Dalton, 134, 199n

Hope Diamond, 186n

Hopkins, Katie, 114

Hopper, Grace Murray, 31, *31*

hornet pupae, as food, 154

"House of the Future," Disneyland, *98*,
    98–9

Huaynaputina volcano, eruption of
    (1600), 70

Huerta, Victoriano, 114, 197n

Hugo, Victor, 71, 178

Huguenots, 70

human genome, mapping of, 4

Hume, David, 15

hunter-gatherers, diet of, 159–60

Huyghen van Linschoten, John, 41

hybridization, 126

Hymenoptera, 139

Illustrated London News, 28

Independent, 154

India, 46, 63, 92, 186n; Adivasis of, 43–4;
    Bengal famine in (1943), 170; Bhopal
    tragedy in (1984), 102; human-insect
    partnerships in, 5, 14; lac, or shellac,
    production in, 14, 39–41, 42–4,
    48–50, 108–9, 196n; mead in, 36; silk
    production in, 59n, 65, 106

indigenous ecological knowledge, 146

indigo *(Indigofera tinctoria)*, 93

Industrial Revolution, 16, 30–1

inheritance: blended, theory of, 127;
    dominant and recessive traits and,
    126; fusion of chromosomal theory
    and Mendel's laws of, 129; Mendel's
    experiments and, 126–8; Morgan's
    Drosophila experiments and, 8, 120,
    122, 124–5, 128–9; resemblances of
    human body and fruit fly and, 9

ink. *See* iron gall ink

insecticides. *See* pesticides and
    insecticides

"insect," origins of word, 18, 20

insects: abundance of, 27; antennae of,
    18–19; chasm between tradition
    and modernity spanned by, 131;
    circulatory system of, 26; compound
    eyes of, 19; entanglement of lives
    of humans and, 14–16, 38, 45,
    58–9, 131–2, 176–7; evolutionary
    success of, 18, 27–8; exoskeletons
    of, 18; as food for humans *(see*
    entomophagy); habitat destruction
    and, 150, 176; human disgust for,
    31–3; human fear of, 3; imagery
    and, 16–18, *17*, 182n; as inspirations
    and exemplars, 16–17; legs of, 18;
    life stages of, 24–5; misguided
    commitment to eradication of, 176
    (see also pesticides and insecticides);
    navigation system of, 19; number of
    species of, 27, 184n; small size of,

insects (continued):
26–7; sounds made by, 64, 173–5;
wingless, 19; wings of, 19
Insects, Food, and Ecology (Brues), 167
Insects and History (Cloudsley-
Thompson), 28–9
"Insects Are Winning, The" (DuPuy), 176
interstellar exploration, entomophagy
in, 171
invasive species, habitats destroyed by,
176
Irish potato famine, 28
iron gall ink, 7, 8, 37–8, 42, 185n
Iroquois Confederacy, 146, 165
Italy, 120; silk production in, 66, 69

Jacobs, Priscilla, 58
jalap (*Ipomoea purga*), 82
Jansen, Zacharias and Hans, 20–1
Japan, 16, 29, 41, 51, 183n, 204n; aesthetic
appeal of "silkworm-moth
eyebrows" in, 56–7; bug concerts
(*mushi-kiki*) in, 174; chemical
poisonings in, 101, 103; dragonfly
motifs in, 16, 17; fleas weaponized
by, 29; insect delicacies in, 161, 162;
meat-eating prohibitions in, 156,
157–8; ramen in, 169; silk in, 56, 74,
76, 77, 105–6
jazz, 39, 50, 175, 187n
Jefferson, Thomas, 157
Jesuits, 93
Jetsons, The, 98
Jia Yi, 62
Jiminy Cricket, 33, 34
Jimmu Tennō, Japanese emperor, 16
Johnston, Sir Harry, 31–2
John the Baptist, 159
Jolie, Angelina, 154
Jordan, Karl, 136
Joubert de la Motte, René-Nicolas, 85
Juan Fernández Islands, 147
jukeboxes, 51, 53–4

June bugs (genus *Phyllophaga*), 26
Justinian, Eastern Roman emperor, 5, 66

Kafka, Franz, 112–13
Kakuta Tôru, 158
Kerckhoff Marine Laboratory, Corona del
Mar, Calif., 123
kermes (*Kermes vermilio*), 90
Kerr, James, 44
Ibn Khallikān, 30
Khotan, sericulture spread to, 65–6
K'iche' people, 29
Kingsolver, Barbara, 175–6
Kobayashi Issa, 76
Kohler, Robert, 125
Korea, silkworm delicacy in (*beondegi*),
161
Kumeyaay people, 167
Kunitz, Stanley, 174–5

lac bugs (*Kerria lacca*), 4, 14, 39–40, 43,
44, 51, 86–8, 92, 130, 186n, 187n; host
plants for, 40, 45, 163; life cycle of,
40, 45; reared by rural women in
India, 108–9. *See also* shellac
"lacquer," origin and use of word, 44
lantana shrub (*Lantana camara*), 138
Lappé, Frances Moore, 168
Larrey, Dominique Jean, 30
larvae, 18, 25n, 26, 27, 37, 92, 161;
battlefield wounds and, 30; crops
eaten by, 148–9; as edible delicacies,
36, 154, 155–6, 158–9, 162, 164–5, 166.
See also silkworms
Lash, Becky, 6
latex, 105
Lawry, James V., 77
leaf-cutter ants (genera *Atta* and
Acromyrmex), 111, 197n
leafhoppers, 30
Le Corbusier (Charles-Édouard
Jeanneret), 97
Leizu, Chinese empress, 60, 61

Leslie, Eliza, 155
Lévi-Strauss, Claude, 36, 185n
lice, 3, 15, 28; rhetoric of genocide and, 32
limitless growth, ideologies of, 103–4
Lindsay, Sarah, 132
Lintner, Joseph Albert, 130
locusts, 3, 27, 153; as food, 153, 159, 167
London, Great Smog of 1952 in, 101
Lorenz, Edward N., 6, 181n
Louis XIV, king of France, 70, 79, 186n
Louis XVI, king of France, 84, 157
Louis-Napoléon Bonaparte, Prince, 74
Lou Shou, 64, 189n
Love Canal, N.Y., 102

Macbeth (Shakespeare), 91
MacBride, Ernest William, 130–1
MacKinnon, J. B., 116
Macready, William, 91
Madagascar star orchid *(Angraecum
    sesquipedale),* 134–7, *135*
Maeterlinck, Maurice, 32
maggots, 24; in Sardinian delicacy, 158–9
magic, boundary between science and,
    131
maguey caterpillar *(Aegiale hesperiaris),*
    or *meocuili,* 165
Mahabharata, The, 39–40
mahua flowers, 43–4
malaria, 3, 24, 29, 89
Malpighi, Marcello, 21
Manchester, England, bee symbolism
    in, 16
Manchuria, weaponized fleas dropped
    on, 29
Mandeville, Bernard, 17
Man Eating Bugs (D'Aluisio), 154
manicure glosses, shellac in, 54
Man of the Woods (Timberlake), 154
"Man's Synthetic Future" (Adams), 97
manure recycling, 71–2
Marie Antoinette, queen of France, 84,
    157

Marine Biological Laboratory (MBL),
    Woods Hole, Mass., *121,* 121–2,
    123–4
Marquis, Don, 113, *113*
Mars expedition, insects as food for, 171
Marston, John, 37
Martianus Capella, 37–8
Massachusetts Institute of Technology,
    98–9
Matassa, Cosima, 51
Mauna Kea volcano, 27, *28*
mauve pigment, 93
Mayas, 29, 35
May bugs, sautéed larvae of, 155–6
McNeill, John, 29
McPhee, John, 61
mead, 36
mealworms, 4, 68, 153, 171
meat: ecological benefits of raising
    edible insects vs., 153–4; Japan's
    prohibitions against eating of, 156,
    157–8; pathogens in production
    facilities and, 167–8; in Stone Age
    diet, 160
Mechanism of Mendelian Heredity, The
    (Sturtevant, Bridges, and Muller),
    129
Meleager of Gadara, 173–4
Memoirs of Military Surgery (Larrey), 30
Mendel, Gregor, 126–8
Mendes, Chico, 174
Menzel, Peter, 154
"Mercy Mercy Me (The Ecology)," 102–3
Merian, Dorothea, 21–4
Merian, Maria Sibylla, 7, 21–5, *25,* 182n,
    183n
Metamorphosis, The (Kafka), 112–13
*Metamorphosis insectorum
    Surinamensium* (Merian), 24–5
methylmercury poisonings, 101
"Metropolitan Nightmare" (Benét), 115
Mexican harlequin bug *(Murgantia
    histrionica),* 20

Mexican Revolution, "La Cucaracha" and, 113–14, 197n

Mexico: *china poblana* silk dress of, 67, 68–9; cochineal production in, 90; Thiéry's theft of cochineal bugs from, 78–85; traditions of entomophagy in, 164–5

mezcal liquor, 165

microscope, invention of, 20–1

Middlesex (Eugenides), 61

Milne, A. A., 37

Minamata, Japan, poisonings in, 101

Misérables, Les (Hugo), 71

Modern Mechanix Magazine, 32

Moffett, Mark, 27

monarch butterfly *(Danaus plexippus),* 6, 18–19, 123n, 175–6

Monbiot, George, 149

Mongol Empire, 67

Monsanto, 98–9

Montezuma II, Aztec emperor, 86

mopane emperor moth *(Gonimbrasia belina),* 162–4, *163*

Morgan, Edmund S., 70

Morgan, Lilian Sampson, 122, 123, 124

Morgan, Thomas Hunt, 8, 120–9, *121,* 130, 132, *133; Drosophila* experiments of, 8, 120, 122, 124–5, 128–9; iconoclasm of, 128–9; Mendel's laws of inheritance and, 126–8, 129; Nobel Prize won by, 122–3; at Woods Hole MBL, *121,* 121–2, 123

Mormons, 58

Morren, Charles, 185n

Morris, Brian, 167

Mosquito, The (Winegard), 29

Mosquito Empires (McNeill), 29

mosquitoes: diseases carried by, 3, 24, 28–9; wingbeat identification of, 173

moths, 137, 159; aesthetic appeal of "moth-feeler eyebrows" and, 56–7; butterflies vs., 56n; cactoblastis *(Cactoblastis cactorum),* 92; clothes,

26; gypsy *(Lymantria dispar),* 75, 92; mopane emperor *(Gonimbrasia belina),* 162–4, *163;* Pandora *(Coloradia pandora),* 166; sphinx *(Xanthopan morganii praedicta),* 136–7. *See also* silkworms

movies, insects as portrayed in, 32–4, *33*

Mughal Empire, 7, 41, 43n

Muir, John, 150–1

Mukasonga, Scholastique, 114

mulberry, white *(Morus alba),* 61, 64–5, 66, 69, 70, 71, 72, 74, 189n, 190n

Muller, Hermann Joseph, 129, 132

Murphy, Michelle, 103

Museum of International Folk Art, Santa Fe, N.Mex., 94

mushi-kiki (bug concerts), 174

music: insect sounds and, 174–5. *See also* records

musical instruments, shellac on, 42, 107

MythBusters, 111

Nabob, The (Foote), 43

Nagasaki bombing (1945), 109–10

Nahuatl people, 146

nam prik mang da (Thai relish), 162

Napoléon Bonaparte, 18, 29–30, 74, 182n

Nash, Linda, 103

National Committee for a Sane Nuclear Policy, 110

National Institutes of Health (NIH), 101

Native Americans, 81, 83, 89, 150; California genocide of 1846–73 and, 32; cloth-coloring techniques of, 92–3; entomophagy traditions of, 165–6, 167; Merian's preservation of terminology of, 24; potatoes cultivated by, 156

Nazi regime, 32, 143–4

Nelson, Dorothy A., 160

neonicotinoids, 148–9

New England: shellac in, 45–6; silk production in, 70–1, 74–5

New Spain: Chinese trade with, 67–9; cochineal production in, 86, 87–9; Florentine Codex and, 89, 165, 192n; potatoes sent to Europe from, 156; Thiéry's theft of cochineal bugs from, 5, 78–85, 91–2. *See also* Mexico

New Testament, 38

New York *Evening Sun,* 113, *113*

New York Times, 52, 110, 147, 204n

Nobel Prize, 9, 122–3, 133, 138

non-timber forest products (NTFPs), 109

Noodle Narratives, The (Gewertz, Errington, and Fujikura), 169

nopal, or prickly pear, cactus (*Opuntia* species), 92, 104; cochineal insects sustained and sheltered on, 83–4, 85, 86, *87,* 87–8

nuclear apocalypse, 197n; cockroach hardiness and, 109–10, 111, 114, 115

nylon, 8, 77

Nymphalidae, 19. *See also* butterflies

Nyuserre Ini, Egyptian pharaoh, 34

Obama, Barack, 53

O'Bannon, Dan, 33

Ó Gráda Cormac, 170n

O'Hara, Charles, 13, 14, 38

Oliver, Mary, 138

Onondaga people, 165

Opuntia. See nopal, or prickly pear, cactus

orchids, pollination of, 23–4, 131–7, 135, 146, 185n

Osterwalder, Elena, 95

Our Stolen Future (Dumanoski and Myers), 103

Our Synthetic Environent (Bookchin), 195n

outer space, fruit flies launched into, 171

Owen, David, 107

Oxford University, 105

Pacific Phonograph Company, 53

painting, cochineal dye in, 90

Palacio, Don Fernán, 82

Palais Royale Saloon, San Francisco, 53

Paleo Diet, The (Cordain), 159–60

Palin, Sarah, 133

Pandora moth *(Coloradia pandora),* 166

Pangaea, 90

Paracas culture, 85

Paracelsus, 100

Pará rubber tree *(Hevea brasiliensis),* 105

parasites: *Alien* films and, 33; Colony Collapse Disorder and, 148, 149; insect-borne maladies and, 29

Paris Academy of Sciences, 85

Parmentier, Antoine-Augustin, 157

Parthian Empire, 62–3

Pasteur, Louis, 178

pathogens, 176; in animal production facilities, 167–8; Colony Collapse Disorder and, 148, 149

pea plants *(Pisum sativum),* 126–7

Peck, Samuel, 46

Pepys, Samuel, 173

Perkin, William Henry, 93

Persip, Charli, 39

pesticides and insecticides, 3, 114, 148–9; cockroaches with immunity to, 114–15; Colony Collapse Disorder and, 148–9; dangers of, 99, 101, 102, 116, 176, 195n

petroleum fly *(Helaeomyia petrolei),* 27

Petrusich, Amanda, 107

pharmaceuticals, 4, 34–5, 107, 108, 168

phonographs, 39, 48, 51. *See also* records

photographic cases, shellac in, 46, *47*

Pilgrim at Tinker Creek (Dillard), 6–7

Pinocchio, 33, 34

Pixar Animation Studios, 33–4, 110

plant-hoppers, 30

Plasmodium parasites, 29

plastics, 45, 98, 99; shellac and, 52, 54

Plato, 20

Pliny the Elder, 20, 63

Plutarch, 34, 63, 188n

poisons, 100; excreted by insects, 87; revelations of environmental toxicology and, 99, 100–3. *See also* pesticides and insecticides

pollen, 137, 138

pollinators, 4, 9, 34, 134–51, 177; coevolution of plants and, 135, 137; environmental advocacy and, 149–50; flight pathways of, 138; floral colors and, 138; indigenous ecological knowledge about, 146; for orchids, 23–4, 134–7, *135*, 146, 185n; pollination process and, 137; reciprocal interaction between plant and, 137. *See also* bees; honeybees

pollutants: airborne, 101; chemical, 102

Polo, Marco, 67

polymer synthesis: silk substitutes and, 77, 99–100; vinyl records and, 52, 54. *See also* Synthetic Age

Popol Vuh, or Mayan book of creation, 29

Popular Science, 104

pork: inefficient to produce, 171; Japanese meat-eating prohibitions and, 158

potato *(Solanum tuberosum),* Europeans' prejudice against, 156–7

poultry production: inefficiency of, 171; pathogens in, 167–8

Procopius, 66

Punesvara, 65–6

Pyle, Robert Michael, 150

Qur'an, 16

ramen, 169

Ransdell Act, 100–1

Rats, Lice and History (Zinsser), 28

Raynal, Guillaume Thomas François, 79

records, 187n; jukeboxes and, 51, 53–4; phonographs and, 39, 48, 51; shellac, 13, 14, 38, 39, 47–8, 50–2, 107; vinyl, 52, 54, 175n

recycling: of shellac, 51–2; silkworm cultivation and, 71–2

red colorants. *See* cochineal dye; colorants

Redzepi, René, 154

Rees, Martin, 76

Reign of Law, The (Campbell), 135

Renoir, Auguste, 90

Revkin, Andrew, 174

rhythm, insect sounds and, 175

rice, 156, 168–9, 170, 189n

Richthofen, Baron Ferdinand von, 66–7

Riley, Charles Valentine, 153

Rimsky-Korsakov, Nikolai, 7, 175

Road to Abundance, The (Rosin and Eastman), 97

Robinson Crusoe (Defoe), 147

Rocky Mountain locusts *(Melanoplus spretus),* 153

Rodgers, Mark, 111

Roger II, king of Sicily, 66

Rogers, Woodes, 147

Romans, Roman Empire, 37, 62–4

Romero, Matías, 93

Root, Amos Ives, 148n

Rosbash, Michael, 9, 132–3

Rosin, Jacob, 97

Rothenberg, David, 175

Rothschild, Walter, 136

Roxburgh, William, 44

Royal George, HMS, 72–3, *73*

rubber, synthesizing substitute for, 104, 105

Rubens, Peter Paul, 90

Rwandan genocide (1994), 114

Safire, William, 43

Sahagún, Bernardino de, 89, 165, 192n

Saint-Domingue (later Haiti), 5, 80, 84–5

Salmasius, Claudius, 44

samurai, dragonfly symbolism and, 16, *17*

Sardinian *casu marzu,* 158–9

saris, silk, 106

Schurz, William Lytle, 68

Science, 97, 124, 138

Scudder, Samuel Hubbard, 111

Seabees (U.S. Navy Construction Batallion), 17

Second Crusade, 66

Selkirk, Alexander, 147

Sen, Amartya, 170

Senate Foreign Relations Committee, 97–8

Seneca people, 146

sericin, 106

sericulture, 104, 188n, 189n; in Britain and its colonies, 70–1, 189n; Chinese origins of, 61–2; climatic anomalies and, 69; contemporary, 105–6; harvesting and processing silkworm cocoons and, 57–8, 59–60; smuggled out of China, 5, 65–6; Trouvelot's failed attempt at, 74–5

"Sex-Limited Inheritance in *Drosophila*" (Morgan), 124

Shakespeare, William, 38, 71, 91

shellac, 4, 8, 13–14, 39–55, 97, 99, 104, 106–9, 177; in beauty products, 54, 55, 108; convenience of, 41; Indian healers' and artisans' uses for, 40–1; jukeboxes and, 51, 53–4; musical instruments and, 42, 107; origin of word "lac" and, 39–40, 44; photographic cases and, 46, 47; production of, 40, 48–50, 51; recording industry and, 13, 14, 38, 39, 47–8, 50–2, 107 (*see also* records); red color of, 41, 44, 47; resurgence of, 38, 54–5, 106–8, 116; trade networks and, 42–3; as varnish, 41, 42, 44, 45, 46–7, 54, 55, 108; in vernacular of U.S. politics, 52–3; vinyl as substitute for, 8, 52, 54, 175n; wartime shortages of, 51–2; as waterproofing material, 41, 51, 55, 107. *See also* lac bugs

"shellacking," as colloquialism, 53

shield bugs, 30

Shoshone people, 166

Silent Spring (Carson), 101–2, 149

silk, 4, 8, 13–14, 38, 41, 45, 56–77, 97, 99–100, 105, 177, 189–90n; Asian trade with New Spain and, 67–9; dresses made of *(china poblana* and *qipao),* 67, 68–9; fashion and, 73–4; life-saving potential of, 106; literary references to, 61, 64, 76, 189n; as medium of exchange, 62; planetary concerns and, 76; resurgence of, 38, 106, 116; Romans' encounters with, 62–4; sumptuary laws and, 74; suppleness, strength, and resilience of, 60, 72–3; synthetic substitutes for, 8, 77, 99–100, 104, 105; weaving techniques and, 66; wild, 61–2, 188n; World War II and, 76–7. *See also* sericulture

silk moths, giant *(Antheraea polyphemus),* 74–5; crossbred with gypsy moths, 75, 92

"Silk Parachute" (McPhee), 61

Silk Roads, 67

silkworms *(Bombyx mori),* 4, 5, 13, 14, 56–61, 57, 76, 112, 189n; aesthetic appeal of "silkworm-moth eyebrows" and, 56–7; beauty products made from cocoons of, 106; caring for, 58–9, 64–5; chrysalids of, in Korean *beondegi,* 161; developmental phases of, 57; eating and soundscapes of, 64–5; harvesting and processing cocoons of, 57–8, 59–60; host plants for, 40, 45 (*see also* mulberry, white); intimacy between humans and, 58–9, 64; waste management and, 71–2. *See also* sericulture

Sima Qian, 190n

Six Flags Great America, Gurnee, Ill., 152–3

slavery, 22, 23, 36, 53, 80, 112, 150, 183n, 197n
sleep cycles, *Drosophila* genes and, 132–3
Smith, Alexander McCall, 163
Smith, John, 7, 70, 112
Smithsonian Folkways Recordings, 175n
Solanaceae (nightshades), 156
Solomon, 16
Solon, 34
Souvenirs entomologiques (Fabre), 178
soy protein, in wasteful food chains, 171
sphinx moth *(Xanthopan morganii praedicta)*, 136–7
spiders, 14n, 19, 140; Hollywood's portrayals of, 32–3; mobility of, 19; silk produced by, 105, 190n. *See also* tarantulas
Stag Beetle (Dürer), 20, 110
Stanton, Andrew, 33–4
Stein, Sir Aurel, 65
Steiner, Achim, 108
Stiles, Ezra, 70–1
Stoermer, Eugene F., 177n
Stradivari, Antonio, 42
Stravinsky, Igor, 50
Sturtevant, Alfred, 128–9, 130, 132
sugarcane, 23, 30, 36
Surinam, 183n; Merian's sojourn in, 21–5
Sutton, Walter, 129
Swammerdam, Jan, 25n
sweeteners, 36. *See also* honey
Synthetic Age, 8, 93, 96–116; Atomic Age and, 109–10; Disneyland's "House of the Future" and, *98*, 98–9; environmental toxicology and, 99, 100–3; insecticides and, 148–9; naturally produced materials too complex to be synthesized and, 99–100; preference for artificial materials and, 97–9

tarantulas, 32, 154; goliath birdeaters *(Theraphosa blondi)*, 23, 183n

Tavernier, Jean-Baptiste, 7, 41, 186n
temperature: below zero, insects that can withstand, 7; cricket's chirps and, 175; weather phenomena and, 69
tendu leaves, 43, 44
Tenmu, Japanese emperor, 157–8
termites, 3, 26, 115, 159
Thailand, giant water bug *(Lethocerus indicus)* eaten in, 161–2
thalidomide, 101
Thiéry de Menonville, Nicolas-Joseph, 5, 78–85, 91–2, 95
Thoreau, Henry David, 176
ticks, 28
Timberlake, Justin, 154
Tlaxcala, Mexico, cochineal in, 88–9
Tokugawa Yoshinobu, 158
Toltec culture, 86
Tomorrowland, Disneyland, *98*, 98–9
Tomson, Robert, 91
Totonaco people, 146
"Touch Me" (Kunitz), 174–5
tournoiement de Turner, 140
toxicology, environmental, 100–3
Trautwein, Michelle, 14
treehoppers, 175
Trouvelot, Étienne Léopold, 74–6, 92
Turner, Charles Henry, 138–41, *140*
typhus and typhoid, 29

Union Carbide, 102
"Union cases," 46, *47*
Unit 731, 29
United Nations, 168; Environment Programme (UNEP), 108; Food and Agriculture Organization, 154
urban food production, 169

Van Gogh, Vincent, 38, 90
van Huis, Arnold, 168
vanilla, 84; artificial versions of, 104, 105
vanilla orchids *(Vanilla planifola)*, 105, 185n

varnishes, 44; shellac as, 41, 42, 44, 45, 46–7, 54, 55, 108
Varroa destructor, 148
Vázquez de Espinosa, Antonio, 88–9
Veracruz, Mexico, *81*; Thiéry's smuggling of cochineal bugs from, 81–3, 84, 85
Versace, Allegra, 106
Villalpando, Cristóbal de, 90
vinyl, 8, 52, 54, 175n
Virginia Colony, 7, 70, 112
Volkamer, Johann Georg, 24
Voltaire, 90
Voyage à Guaxaca (Thiéry), 78

Wallace, Alfred Russel, 135–6
WALL-E, 110
Walter, Bruno, 50
Wanderer, 112, 197n
Wanli, Ming emperor, 68
Ward, Alice Bailey, 75
Washington, George, 14–15
wasps, 29, 33, 37, 137, 139
water bug, giant *(Lethocerus indicus),* 161–2
Watt, Sir George, 43
Wax or the Discovery of Television Among the Bees, 19n
weaponized insects, 29
weather: global interconnectedness and, 69; insect messages about changes in, 175–6. *See also* climate change
weevils, 27
Weiner, Jonathan, 125
wēkiu bug *(Nysius wekiuicola),* 27, 28
Westwood, John Obadiah, 27
wheat, 156, 168, 169

Whitman, Walt, 133
Why Not Eat Insects? (Holt), 155, 172
Wilde, Oscar, 73–4, 190n
Wilkes, John, 96
Wilson, Edmund Beecher, 122
Wilson, E. O., 9–10, 22
Winegard, Timothy C., 29
Winnie-the-Pooh, 37
Wöhler, Friedrich, 96
Woman of No Importance, A (Wilde), 73–4
Wood, Thomas William, *135, 136*
Woods Hole, Mass., *102,* 130; Children's School of Science in, 6, 123n; MBL in, *121,* 121–2, 123–4
Woodworth, Charles William, 8
World Health Organization, 29
World War II, 51–2, 61, 76–7, 105
World War III, 110

Xerces Society for Invertebrate Conservation, 150
Xiliang, General, 72
Xiongnu, 62
X-rays, genetic mutations caused by, 132

Yangshao, 61
yellow fever, 3, 29, 89
Yolngu people, 35–6, 146
Young, Michael W., 9, 133

zazamushi (Japanese insect delicacy), 162
Zika virus, 3, 29
Zinsser, William, 28, 46–7, 107
Zuk, Marlene, 160

ILLUSTRATION CREDITS

Page

x   Map by Nick Springer, Springer Cartographics, 2019

17   Courtesy of the Metropolitan Museum of Art

28   Courtesy of the University of Hawai'i Institute for Astronomy

31   Courtesy of Vassar College Archives and Special Collections

49   Courtesy of Kurt Starlit

57   Courtesy of the Commonwealth Scientific and Industrial Research Organisation, Australia

60   Courtesy of the Suzhou Silk Museum, Chengdu, China

73   Courtesy of the National Maritime Museum, Greenwich, London

87   Courtesy of the Newberry Library

98   Courtesy of the United States Library of Congress

102   Courtesy of the IEEE Oceanic Engineering Society

121   Courtesy Marine Biological Laboratory Archives

125   Courtesy of Sanjay Acharya

135   Courtesy Darwin Correspondence Project, Cambridge University

140   Courtesy of Charles I. Abramson

143   Courtesy the Nobel Foundation Archive

163   Courtesy AP Photo/Tsvangirayi Mukwazhi

Edward D. Melillo is the author of *Strangers on Familiar Soil: Rediscovering the Chile-California Connection,* which won the Western History Association's 2016 Caughey Prize for the most distinguished book on the American West. He is professor of history and environmental studies at Amherst College, where he teaches courses on global environmental history, the history of the Pacific world, nineteenth-century U.S. history, and commodities in world historical perspective. For the duration of a 2017–18 Mellon New Directions Fellowship, he studied *ʻōlelo Hawaiʻi* (Hawaiian language) on Oʻahu. Melillo received his Ph.D. and his M.Phil. from Yale University and his B.A. from Swarthmore College. He lives with his son in South Hadley, Massachusetts.

A NOTE ON THE TYPE

This book was set in Minion, a typeface produced by the Adobe Corporation specifically for the Macintosh personal computer, and released in 1990. Designed by Robert Slimbach, Minion combines the classic characteristics of old-style faces with the full complement of weights required for modern typesetting.

Composed by North Market Street Graphics, Lancaster, Pennsylvania

Printed and bound by Berryville Graphics, Berryville, Virginia

Designed by Maggie Hinders